Salt Crystallization in Porous Media

SCIENCES

Mechanics, Field Director – Gilles Pijaudier-Cabot

Geomechanics, Subject Head – Cino Viggiani

Salt Crystallization in Porous Media

Coordinated by

Hannelore Derluyn
Marc Prat

iSTE

WILEY

First published 2024 in Great Britain and the United States by ISTE Ltd and John Wiley & Sons, Inc.

ISTE Ltd
27-37 St George's Road
London SW19 4EU
UK

www.iste.co.uk

John Wiley & Sons, Inc.
111 River Street
Hoboken, NJ 07030
USA

www.wiley.com

Library of Congress Control Number: 2023942766

British Library Cataloguing-in-Publication Data
A CIP record for this book is available from the British Library
ISBN 978-1-78945-114-6

ERC code:
PE8 Products and Processes Engineering
PE8_3 Civil engineering, architecture, maritime/hydraulic engineering, geotechnics, waste treatment
PE10 Earth System Science
PE10_11 Geochemistry, crystal chemistry, isotope geochemistry, thermodynamics

Contents

Preface

Hannelore DERLUYN[1,2] **and Marc PRAT**[3]

[1] *LFCR, CNRS, Université de Pau et des Pays de l'Adour, France*

[2] *DMEX, CNRS, Université de Pau et des Pays de l'Adour, France*

[3] *IMFT, CNRS, Université de Toulouse, France*

Salt crystallization phenomena in porous media are found in important applications such as soil salinization, the underground storage of CO_2 in saline aquifers or the preservation and conservation of built heritage. In connection with this last application, the crystallization of salts in building materials or porous stones of many components in our architectural or cultural heritage is considered to be one of the most severe threats to their integrity.

The stresses induced by crystallization in pores can indeed lead to fracturing, flaking, or even honeycombing, seriously altering the esthetics of the elements concerned, or even their structural stability. Crystallization in soils can also lead to structural disorders when induced swelling phenomena affect the foundations or surface structures such as roads, tracks and other paved surfaces. With these applications in sight, this book provides an overview of the state of the art of scientific knowledge concerning the crystallization of salt in porous media.

An important feature of this area of research is its interdisciplinary character. For example, when crystallization and the damage induced result from the evaporation of the solution contained in the pores of a material, the analysis involves, in addition to the evaporation phenomenon in a porous medium itself, physicochemical phenomena. These include precipitation or dissolution, solute transport phenomena inside the pore space, capillarity and viscous flow phenomena, thermal, and mechanical phenomena (fracturing, swelling, modification of the pore structure,

Salt Crystallization in Porous Media,
coordinated by Hannelore DERLUYN and Marc PRAT.

etc.). Besides the great variety of phenomena, what makes it more complex is that these different phenomena are generally coupled together so that the most complex situations require the consideration of these couplings.

An immediate consequence is that in reality different communities of researchers are interested in this field of research. Each bringing their own methodologies and areas of interest, several lines of research have emerged over the years. One part concerns thermodynamics, with a description of equilibrium situations, which can quickly become very complex when several salts are present; this is frequent in applications and also because of metastable states, which are often found. The description of equilibrium states, phase diagrams and the impact of confinement, especially in pores of submicronic size down to just a few nanometers, can alone occupy a researcher's life. If thermodynamics is the key to describing these often subtle equilibria, it is generally not sufficient on its own to comprehend a practical situation of crystallization in a porous medium given the importance of transport phenomena. Very often, the growth kinetics of a developing salt structure, whether growing on the surface of the porous material (efflorescence) or within it (subflorescence), can only be described correctly through the sufficient understanding of the phenomena controlling the transport of ions within the porous material as well as within the salt structure itself, the latter often being porous (case of NaCl in particular). Thermodynamics is still very present in describing the local equilibrium at interfaces, but using methods of fluid mechanics and transport phenomena in porous media then becomes essential. Finally, poromechanics comes into play when it comes to analyzing situations leading to fractures and/or deformations. As evidenced by this book, in reality the combined consideration of all these aspects is still only imperfectly carried out. For example, work focusing on the transport and growth laws of efflorescence and subflorescence has mainly focused on the simplest cases where only one salt is present in the solution. Similarly, some aspects highlighted by studies focusing on the transport and growth of salt structures are not taken into account or only in an oversimplified way in the most advanced models of poromechanics.

These elements indicate that the field of salt crystallization remains a very open field of research where much remains to be uncovered. Also, the specificities of this field of research are found in the chapters themselves, which develop in a relatively autonomous way.

In this context, the introduction takes up the previous elements by placing them in relation to the literature. The book is then divided into two main parts. In the first part, the objective is to focus on the basic principles that should be mastered to study problems related to the crystallization of salts in porous media. So, Chapter 1 focuses on the thermodynamic properties of saline solutions. The key concept of crystallization pressure is the subject of Chapter 2. The main results concerning the aspects of solute transport in porous media and growth of salt structures are presented in Chapter 3. The state of the art concerning the theories of poromechanics for the prediction of salt-related damage is proposed in Chapter 4. Chapter 5 presents experimental observations in geomaterials, mainly from modern imaging techniques.

Part 2 is more applicative and illustrates not only the operational nature of the fundamental concepts in the previous chapters, but also the developments still to be made to understand and model certain real scenarios. Chapter 6 presents situations where crystallization in the underlying terrain has a considerable impact on engineering structures, bridges or tunnels. It can also be seen as a transition chapter because the coupled thermohydromechanical and chemical modeling of salt precipitation and the deformation of rocks in contact with infrastructures is also presented. This model provides good results compared to field observations. Chapter 7 presents a very important field of application, that of the alteration of porous media due to the crystallization of salts in the context of built heritage, from identifying the origin of salts and the degradation mechanisms to conservation and restoration techniques. Finally, Chapter 8, in continuity with Chapter 7 and the issue of the preservation of built heritage, emphasizes the impact of environmental conditions and the evolution of climatic factors. Like Chapter 7, it illustrates the complexity that can be encountered in real-life scenarios when multiple salts are present.

These chapters have been prepared by specialists in each field concerned. Their name and affiliation are indicated at the beginning of the chapter and at the end of the book. We thank them very warmly for their involvement in the creation of this book.

The latter is aimed primarily at students at master's degree level and above and at researchers and engineers interested in this field of research and wishing to have an overview of its various facets. However, this book may be of interest to a wider audience, whether those more directly concerned with the applications or those responsible for teaching, certain parts being able to lend themselves to the illustration of basic concepts beyond the problem of salt itself.

Finally, this book would probably not have seen the light of day without a certain structuring of the community of researchers working on the crystallization of salts in porous media, in particular via the series of CRYSPOM workshops "Crystallization in Porous Media", which, since 2008, at the initiative of Noushine Shahidzadeh and the late Olivier Coussy, allows regular exchanges on this topic.

As with most research the results presented in the various chapters could not have been obtained without the contribution of many people, starting of course with the various doctoral students who developed their thesis on these topics, and the exchanges or collaborations with many colleagues. It is impossible to name them all. May they accept our deepest gratitude.

Introduction

Marc PRAT
IMFT, CNRS, Université de Toulouse, France

The crystallization of a salt in a porous medium is studied in relation to different applications. First, because of its destructive effect. The latter is easily observable when visiting many monuments and historical sites. As noted in the classic work by Goudie and Viles (1997), this is the case, for example, of the Sphinx in Egypt, the stone monuments of Petra in Jordan, several cathedrals in Mediterranean regions, or even the city of Venice. As discussed in Chapters 7 and 8 of this book, the attack by salts results in different forms of degradation such as cracking, flaking, honeycomb weathering (Figure I.1) and can go as far as pulverizing the material. All of the processes linked to the crystallization of salts that lead to the degradation of porous rocks are defined under the term “haloclasty”. Thus, haloclasty is considered to be the main phenomenon causing the alteration of historical monuments in the Mediterranean region.

Although haloclasty has been observed since antiquity (Goudie and Viles 1997), it is only since the 19th century that these phenomena began to be studied seriously. For example, Brard (1828) presented research where the crystallization of salt was studied as analogous to the effect of freezing on porous stones, whereas initial discussions, aiming at explaining that the growth of a crystal in confinement generates constraints, date back to 1853 with the work of Lavalle (1853). Haloclasty has also been identified as an important phenomenon in the evolution of landscapes, contributing to erosion not only on Earth but possibly also on Mars (Malin 1974).

a) b) c)

Figure I.1. *Examples of damage due to salt crystallization: (a) cracking, (b) flaking (Motomachi Buddha, Oita, Japan) and (c) honeycomb weathering (Santander Cathedral, Spain) (photos (a) and (c), B. Leclère (Leclère 2021); photo (b), H. Derluyn). For a color version of this figure, see www.iste.co.uk/derluyn/salt.zip*

As for many other phenomena, the scientific approach to haloclasty and more generally, to the crystallization of salts in porous media and associated phenomena, gained considerable interest in the 20th century. Among the seminal works often cited, we can mention the work of Correns (1949), who proposed an expression of the crystallization pressure, a central concept in the analysis of the generation of stresses due to crystallization (see Chapter 2 of this book). This formulation has since been revisited and we can refer to Steiger (2005) for a more recent presentation according to a purely thermodynamic equilibrium approach. More recent studies (e.g. Gagliardi and Pierre-Louis 2019) indicate, however, that it is important to consider non-equilibrium effects when analyzing the force exerted by a crystal in a pore. The subject of stress generation, due to the crystallization of salt in pores (Scherer 1999; Coussy 2006; Noiriel et al. 2010; Scherer et al. 2014), inducing degradation phenomena in porous stones, as well as the detrimental effects that it can cause on engineering structures (see Chapter 6 of this book), is not the only motivation to study the impact of the presence of one or more salts in a porous medium. Other applications include the following:

– *Soil salinization*: salinization refers to the accumulation of salts in soils at levels toxic to most plants. It is a main cause of desertification, erosion and land degradation. It makes the soil unsuitable for agriculture. Very likely linked to global warming, the phenomenon is accelerating (Litalien and Zeeb 2020) and threatens ever larger areas of land. Irrigation is one of the main human causes of salinization. In the event of excessive irrigation, the soil is wet at depth, which establishes hydraulic continuity causing the salt to rise to the surface. A good understanding of the transport of dissolved salt in an unsaturated porous medium is therefore key to understanding and analyzing this phenomenon.

– *Soil evaporation*: the phenomenon of evaporation in the presence of dissolved salt often leads to the formation of a salt crust on the surface of the soil. Depending on the formation conditions, the formation of the crust can severely reduce evaporation (Fujimaki et al. 2010) and therefore affect soil–atmosphere exchanges. Despite advances (see Chapter 3 of this book), the detailed understanding and modeling of this phenomenon remain to be fully understood.

– *Ground heave*: when crystallization occurs at depth, the soil layers above may heave due to stress caused by crystallization. This mechanism is illustrated in Figure I.2. We find here a mechanical effect in the context of granular media rather than consolidated media such as porous rocks. This uplift phenomenon can lead to major surface disorders (Feng et al. 2020; also see Chapter 6). Although clearly identified (Hird and Bolton 2016), it is still insufficiently described with a lack of predictive models.

– *Underground storage of CO_2*: carbon capture and storage (CCS) in geological formations is a technique aimed at limiting the accumulation of greenhouse gases in the atmosphere. Some of the main issues related to this technique concern the injectivity conditions of the wells used to maintain a high level of injection (Peysson et al. 2014). When this injection takes place in a saline aquifer, salt precipitation can occur, leading to the clogging of pores in the well, a drop in permeability and ultimately the deterioration of injection conditions. A better understanding of these phenomena is therefore desirable to improve the implementation of this technique.

– *Impregnation with a saline solution*: with the Covid-19 pandemic, it has been shown that individual masks are essential tools to protect against airborne viral particles. In this context, hypertonic saline solution has been shown to effectively reduce infectious viral load on treated masks (Tatzber et al. 2021). The method consists of impregnating the mask with a saline solution of NaCl and then drying it. We can refer to Börnhorst et al. (2016) for a discussion on the impregnation of a porous medium by a saline solution in connection with other contexts.

– *The use of crystallization as a cheap method for the recovery of metals*: a simple method to recover metals from mine tailings is to soak them in a saline solution and subject them to evaporation (Cala-Rivera et al. 2018). In this situation, as explained in Chapter 3, a flow is induced toward the surface in the porous medium. This flow transports the species in solution to the surface where the salt crystallizes. The metals become trapped in the efflorescence, which then only needs to be extracted and dissolved to recover the metals in question.

– *Conservation and restoration of built heritage*: considering the damage that crystallization can cause, various techniques have been developed to limit the risk of degradation or restore the affected areas, whether for recent constructions or for the old parts of built or cultural heritage (frescoes, paintings, sculptures, etc). These aspects are discussed in Chapter 7.

– *Design of distillation unit evaporators*: desalinating seawater is an important solution to meet the growing demand for drinking water, but also potentially in the context of the energy transition and the production of hydrogen by electrolysis, which also uses water as a resource. In this context, solar-thermal water desalination is an interesting solution, which can generate clean water without significantly depending on fossil energy. In devices designed for this purpose (Wang et al. 2020), we find (nano)porous media, evaporation and a saline solution, ingredients common to the applications already mentioned.

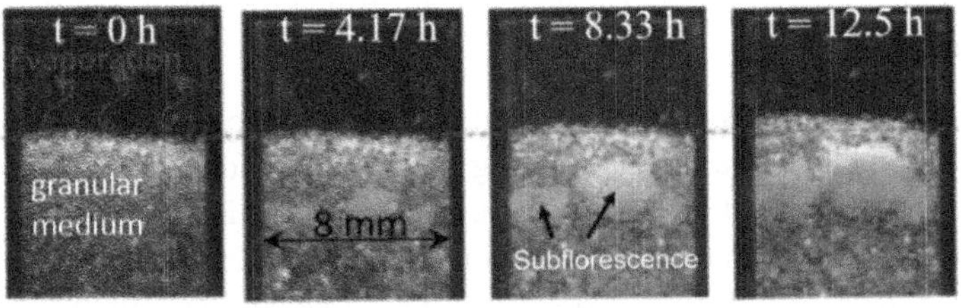

Figure I.2. *Drying a NaCl solution in a Hele–Shaw cell filled with glass beads. The process leads to the formation of subflorescence causing the granular medium to move upwards and lift off the surface (Diouf 2018). For a color version of this figure, see www.iste.co.uk/derluyn/salt.zip*

These different applications illustrate the multidisciplinary nature of the field of research that is the subject of this book. Thus, different communities of researchers contribute to its development including specialists in geophysics, geomorphology, underground or surface hydrology, physicochemistry, physics, chemistry, thermodynamics, civil engineering, transfer in porous media, poromechanics, engineering, or even more directly, in the relevant applications.

Even if each specialist can find their own reasons for their interest in this field of research, it is clear that the analysis of the most complex situations requires a certain level of interdisciplinarity. Consider, for example, the very classic scenario where crystallization results from the evaporation of a saline solution contained in a porous medium. In this scenario, evaporation induces an increase in the salt concentration in the solution within the porous medium by the effect of advection and/or volume reduction (see Chapter 3) until a sufficient concentration is reached for crystallization to occur and subflorescence and/or efflorescence to develop. As shown schematically in Figure I.3, the analysis of this type of situation involves focusing on the transport of dissolved salts in a porous medium subjected to evaporation and often unsaturated (case of drying for example), physicochemical aspects, including the often complex thermodynamic equilibria characterizing the solutions (Chapters 1, 7 and 8), precipitation–dissolution phenomena and the dependence of the properties (viscosity, surface tension, wettability, etc.) on the salt concentration, etc. When crystallization has notable mechanical consequences (cracking, movement of grains, etc.), the preceding phenomena become coupled with mechanics, which is often tackled within the framework of macroscopic poromechanics (Chapters 4 and 6), but also with research at the pore scale (Chapter 2).

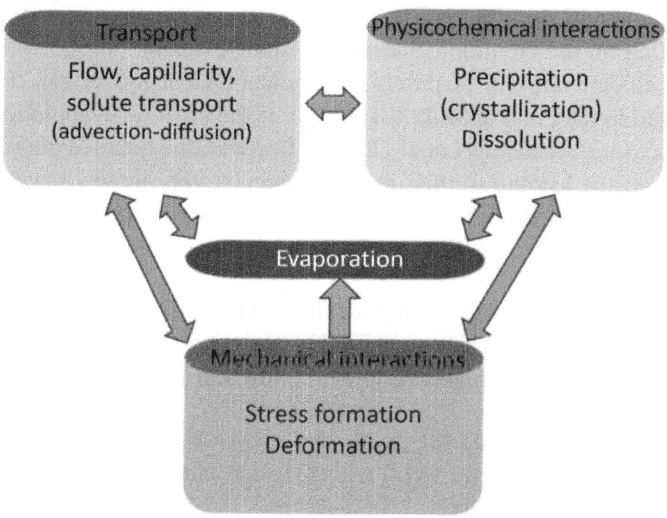

Figure I.3. *Evaporation of a saline solution in a porous medium: simplified diagram of coupled processes. For a color version of this figure, see www.iste.co.uk/derluyn/salt.zip*

Indeed, as is conventional for porous media, the crystallization of salt in porous media and the associated phenomena can be studied at different scales, from the pore scale to the scale of underground formations, passing through the pore network scale and the Darcy scale. The current situation is, however, far from having reached the level of development where the model at a given scale, for example the Darcy scale, is deduced in a rigorous way, that is to say by upscaling methods, from the formulation of the model at a lower scale, a pore scale model typically. Even if there is research going in this direction, the present situation is that each study is developing in a fairly independent way. To give just one example, several works have been dedicated in recent years to the crystallization pressure from experiments at the scale of a pore or a channel (Chapter 2). On the other hand, the concept of crystallization pressure is also central in poromechanical models (Chapter 4). However, how a growing subflorescence within the pore network generates stresses has not yet been properly analyzed. In other words, the intermediate link is not understood, which necessarily introduces some uncertainty on the real applicability of the poromechanical models developed so far.

Even if there are modeling studies where all the phenomena mentioned in Figure I.3 are taken into account (Koniorczyk and Gawin 2012; Derluyn et al. 2014), many phenomena are in fact still poorly understood or poorly described. To cite just one, we can mention the effect of the formation of salt crusts on evaporation, which can, depending on the conditions, severely reduce evaporation or, on the contrary, have no significant impact. A purely phenomenological model has, for example, been proposed to take into account the impact of the crust on evaporation when this greatly reduces it (Grementieri et al. 2017). However, the detailed physical analysis of the mechanisms leading to this impact remains a very largely open question. A recent study suggests that it is due to the formation of submicron pores on the surface of the crust and to the Kelvin effect (Licsandru et al. 2022).

This also explains, beyond the applications, that research still continues, not only by trying to integrate all the phenomena, but also in a specific way. This book, by the structuring of its topics, nicely reflects the different lines of current research.

From a more methodological point of view, research on the study of porous media follows a fairly general trend with the increasingly frequent use of modern imaging techniques (Chapter 5). Access to these techniques (X-ray microtomography, neutron radiography, etc.) renews the questions with the possibility of considering all the complexity, in particular geometric, of real environments at the scale of their microstructure. Nevertheless, given the complexity and diversity of the phenomena associated with crystallization in porous media, the implementation of very diverse approaches, experimental and numerical, including on model systems that can go as far as a simple capillary tube, remains completely of topical interest.

Finally, it should be noted that simple saline solutions, most often made up of a single salt, are considered in the majority of academic studies. This is natural in a logic of increasing complexity, where the simplest situations must first be understood. As illustrated in Chapters 7 and 8, the situation in the field can be much more complex with the presence of many salts. Even in the case of simple solutions, the salt considered leads to specificities. For example, sodium chloride has only one crystalline form; it is anhydrous, under typical temperature conditions in applications, whereas sodium sulfate, for example, can have several crystalline forms. These specificities have important consequences, which imply clearly distinguishing what can be generic, that is to say common to all salts, from what is specific to certain salts. For example, from an application point of view, it is well established that sodium sulfate presents a greater threat than sodium chloride with regard to the destructive effect of salts.

In summary, the study of the crystallization of salts in porous media interests specialists from very diverse disciplines, whether in connection with applications or because of the many facets that this field of research presents. It is rich in open questions and although many advances have been made over the past decades, as evidenced by this book, much remains to be done, particularly with regard to predictive models, in connection with field situations. Nevertheless, this book bears witness to the degree of maturity reached, which, although it does not always provide answers to practical questions, offers a very coherent body of knowledge that also makes it possible to analyze very concrete situations in a convincing manner, as evidenced in Chapter 6.

References

Börnhorst, M., Walzel, P., Rahimi, A., Kharaghani, A., Tsotsas, E., Nestle, N., Besser, A., Kleine Jäger, F., Metzger, T. (2016). Influence of pore structure and impregnation–drying conditions on the solid distribution in porous support materials. *Drying Technology*, 34(16), 1964–1978.

Brard, C.-P. (1828). Sur le Procédé pour reconnaître, immédiatement, les pierres qui ne peuvent pas résister à la gelée et que l'on désigne ordinairement par les noms de pierres gelives ou pierres gelisses. *Annales de chimie et de physique*, 38, 160–192.

Cala-Rivera, V., Arranz-González, J.C., Rodríguez-Gómez, V., Fernández-Naranjo, F.J., Vadillo-Fernández, L. (2018). A preliminary study of the formation of efflorescent sulfate salts in abandoned mining areas with a view to their harvesting and subsequent recovery of copper. *Minerals Engineering*, 129, 37–40.

Correns, C.W. (1949). Growth and dissolution of crystals under linear pressure. *Discussions of the Faraday Society*, 5, 267–271.

Coussy, O. (2006). Deformation and stress from in-pore drying-induced crystallization of salt. *Journal of the Mechanics and Physics of Solids*, 54, 1517–1547.

Derluyn, H., Moonen, P., Carmeliet, J. (2014). Deformation and damage due to drying-induced salt crystallization in porous limestone. *Journal of the Mechanics and Physics of Solids*, 63, 242–255.

Diouf, B. (2018). Évaporation avec cristallisation de sel en milieu granulaire : localisation des cristaux et déplacements mécaniques induits. PhD Thesis, Université de Toulouse, Toulouse.

Grementieri, L., Molari, L., Derluyn, H., Desarnaud, J., Cnudde, V., Shahidzadeh, N., de Miranda, S. (2017). Numerical simulation of salt transport and crystallization in drying Prague sandstone using an experimentally consistent multiphase model. *Building and Environment*, 123, 289–298.

Feng, R., Wu, L., Wang, B. (2020). Numerical simulation for temperature field and salt heave influential depth estimation in sulfate saline soil highway foundations. *International Journal of Geomechanics*, 20(10), Article ID 04020196.

Fujimaki, H., Shimano, T., Inoue, M., Nakane, K. (2006). Effect of a salt crust on evaporation from a bare saline soil. *Vadose Zone Journal*, 5(4), 1246–1256.

Gagliardi, L. and Pierre-Louis, O. (2019). The nonequilibrium crystallization force. *Europhysics Letters*, 127(5), 59002.

Goudie, A. and Viles, H. (1997). *Salt Weathering Hazards*. Wiley, New York.

Hird, R. and Bolton, M.D. (2016). Migration of sodium chloride in dry porous materials. *Proceedings of the Royal Society A*, 472, 20150710.

Koniorczyk, M. and Gawin, D. (2012). Modelling of salt crystallization in building materials with microstructure – Poromechanical approach. *Construction and Building Materials*, 36, 860–873.

Lavalle, J. (1853). Recherche sur la formation lente des cristaux. *Comptes rendus de l'Académie des Sciences*, 36, 493–495.

Leclère, B. (2021). Damage induced by crystallisation in confined reservoir conditions. PhD Thesis, Université de Pau et des Pays de l'Adour, Pau.

Licsandru, G., Noiriel, C., Geoffroy, S., Abou Chakra, A., Duru, P., Prat, M. (2022). Detachment mechanism and reduced evaporation of an evaporative NaCl salt crust. *Scientific Reports*, 12(7473).

Litalien, A. and Zeeb, B. (2020). Curing the earth: A review of anthropogenic soil salinization and plant-based strategies for sustainable mitigation. *Science of the Total Environment*, 698(1), 134235.

Malin, M.C. (1974). Salt weathering on Mars. *Journal of Geophysical Research*, 79(26), 3888–3894.

Noiriel, C., Renard, F., Doan, M.L., Gratier, J.P. (2010). Intense fracturing and fracture sealing induced by mineral growth in porous rocks. *Chemical Geology*, 269, 197–209.

Peysson, Y., André, L., Azaroual, M. (2014). Well injectivity during CO_2 storage operations in deep saline aquifers. Part 1: Experimental investigation of drying effects, salt precipitation and capillary forces. *International Journal of Greenhouse Gas Control*, 22, 291–300.

Scherer, G.W. (1999). Crystallization in pores. *Cement and Concrete Research*, 29, 1347–1358.

Scherer, G.W., Flatt, R.J., Caruso, F., Aguilar Sanchez, A.M. (2014). Chemomechanics of salt damage in stone. *Nature Communications*, 5, 1–16.

Steiger, M. (2005). Crystal growth in porous materials I: The crystallization pressure of large crystals. *Journal of Crystal Growth*, 282(3/4), 455–469.

Tatzber, F., Wonisch, W., Balka, G., Marosi, A., Rusvai, M., Resch, U., Lindschinger, M., Moerkl, S., Cvirn, G. (2021). Coating with hypertonic saline improves virus protection of filtering facepiece manifold – Benefit of salt impregnation in times of pandemic. *International Journal of Environmental Research and Public Health*, 18(14), 7406.

Wang, X., Hsieh, M.L., Bur, J.A., Lin, S.Y., Narayanan, S. (2020). Capillary-driven solar-thermal water desalination using a porous selective absorber. *Materials Today Energy*, 17, 100453.

PART 1

Fundamental Aspects

1

Thermodynamics and Salt Crystallization Kinetics

Lionel MERCURY
ISTO, CNRS, BRGM, Université d'Orléans, France

1.1. Introduction

The crystallization of salts from aqueous solutions is a common, even daily event, the appearance of which does not seem to cause much difficulty. The increase in temperature or the evaporation of water are two typical causes, sometimes concurrent, which cause this phase transition, while the initial composition of the solution makes it favorable to the generation of salts.

Yet it is not trivial to accurately predict when and in what form a salt will appear in water. Also, it is not so easy to know where the salt will form, especially when the solution is moving through a network of channels and/or pores. The calcification of heating pipes, very heterogeneous and often atypical, is a good example of these complexities.

The purpose of this chapter is to look precisely at what thermodynamics, the science behind the driving forces, and kinetics, the science of transformation paths, say about these phase transitions. Far from an overview of commonly reviewed concepts, this chapter aims to highlight some subtle concepts that establish unsuspected connections between the water of the solution, the species which are dissolved therein, the pore walls enclosing it and local physicochemical conditions.

Salt Crystallization in Porous Media,
coordinated by Hannelore DERLUYN and Marc PRAT.

1.2. Thermodynamic driving force

1.2.1. *Equilibrium, spontaneity, irreversibility*

Energy is always difficult to define in a strict sense, but a fairly natural understanding of this concept can be made by pointing out that energy is a product of a force by an extensive term (quantity dependent); typically the elastic energy of a spring is written as: $E = F.l$, with F being the force of the spring and l its length. The force characterizes the intensity under which the considered energy is present in the system, for example, the force of a table lamp spring is not comparable to that of a car shock absorber. The term extensive translates the extent under which the force can be expressed: here, the elongation of the spring allows a "stronger" reservoir, more "intense", to communicate elastic energy to the spring studied. Similarly, during its shortening, the spring makes energy available in proportion to the variation in its length.

According to the first law, the law of conservation of energy, two reservoirs in contact will exchange energy until their energy forces equalize, when all exchange stops. The state of equilibrium is thus the attractor of any system that is not in equilibrium: the evolution of natural systems is neither arbitrary nor random. The thermal energy goes from the hot source to the cold source, and the opposite is never observed: the energy therefore travels in the direction of decreasing temperature, the temperature here measuring the intensity, the force under which the thermal energy is present in the two reservoirs. The first law therefore makes it possible to state that the energy is conserved during the exchanges, which take place from the most intense reservoirs to the less intense reservoirs, by an exchange of extensity that decreases in the strongest reservoir and increases in the weakest reservoir. One can give as an example the flow of electric charges, electric extensity, which moves in the direction of decreasing electric voltage.

The second law adds precision to these possible developments. This law of directionality fixes the direction in which the transformations become spontaneous. This point is often misunderstood, in the sense that the first law already states that the transformations are not arbitrary, since they occur in the direction of decreasing intensity. The point that the second law clarifies is the fact that there are transformations that take place without exchanging extensive terms between two reservoirs. When a hot source heats a cold source, there is a flow of thermal extensity (entropy, immaterial extensity) from the first to the second. It is not a spontaneous transformation, the contact between the two sources must be ensured so that the extensity flows from one to the other. Spontaneity occurs when a transformation takes place without the need for an entropy source: subjected to an intensity gradient, the system begins to create entropy in proportion to this gradient, as if it were in contact with an entropy source. The immaterial aspect of entropy makes it possible to create it without contravening the law of conservation of matter, and since it is an extensity and not an energy, this creation does not contravene the first law either. The system therefore evolves spontaneously, since the necessary entropy is created on site, and irreversible because it created entropy that can no longer be destroyed. To conclude, it is therefore clear that a system subjected to an energy gradient can transform itself even in the absence of any extensive reservoir, if the creation of entropy makes it possible to establish an energy exchange chain that respects the first law.

With these rather abstract considerations having been made, now let us see how a spontaneous transformation can be quantified. For this, it is necessary to know if the creation of entropy is possible for the transformation of interest. One might assume that it suffices to measure the change in entropy in the system. The difficulty is that the entropy exchanged and the entropy created are indistinguishable, while spontaneity is only expressed by the second. The solution lies in monitoring the thermodynamic potential of the system, which is a property whose value changes in direct relation to the creation of entropy. It is easy to understand that an isothermal, isochoric system (constant *T*, *V*) can only create entropy in the same way as an isothermal, isobaric system (constant *T*, *P*); in fact, several thermodynamic properties can become a thermodynamic potential depending on the state variables of the system. In the case that concerns us with nucleation and crystal growth in an aqueous solution, the adequate thermodynamic potential is known as the free enthalpy or Gibbs free energy.

Gibbs free energy, as thermodynamic potential, is written as follows:

$$\Delta G_{T,P} = -T.\Delta S_{created} \quad [1.1]$$

Any decrease in free enthalpy during an isothermal, isobaric transformation is therefore associated with a creation of entropy, reflecting an irreversible and spontaneous transformation. The reverse transformation is in fact impossible.

1.2.2. *Parameters controlling the driving force*

1.2.2.1. *Activity of dissolved species*

As we are interested in chemical transformations, it is appropriate to note that the free enthalpy of one mole of a chemical compound is equal to its chemical potential. However, the chemical potential makes it possible to integrate into the reasoning the characteristic terms of the composition of the system. Whatever the compound and its physical state (liquid, solid, gas, solute), its chemical potential μ is written as follows:

$$\mu = \mu^\circ + RT.\ln(a) \qquad [1.2]$$

where μ° is the chemical potential of the compound when it is in a conventional state, called the standard state; a is the thermodynamic activity of the compound; R is the ideal gas constant and T is the temperature. The activity is equal to 1 in the standard state and deviates from this value when the compound deviates from this conventional state. A standard state has at least four attributes: composition, "aggregation" state, temperature or range of temperatures, and pressure or range of pressures. The most common are written as follows:

– $a_{solid/liquid} = \gamma x$ for solids and liquids ∀ T,P, x being the mole fraction of the solid or liquid in the mixture. So $a_{solid/liquid} = 1$ for pure solids or liquids (in the stoichiometric sense, $x = 1$).

– $a_{gas} = \chi P/\chi^\circ P^\circ$ for gases ∀ T with $\chi^\circ = 1$ and $P^\circ = 1$ bar. The product of the fugacity coefficient and the pressure is the fugacity, denoted as f: $a = f/f^\circ$ with $f^\circ = 1$.

– $a_{solute} = \gamma\ m/\gamma^\circ m^\circ$ for solutes ∀ T,P with $\gamma^\circ = 1$ and $m^\circ = 1$ mol/kg. The standard state thus appears as a hypothetical situation where a solute concentrated at 1 mol·kg (rather high value) would behave like a solute in infinite dilution ($\gamma^\circ = 1$).

– A compositon term (x, P, m) is combined with the activity coefficient (γ, Xi) which enhances (γ, $\chi > 1$) or diminishes (γ, $\chi < 1$) the contribution of the composition to the chemical potential.

The case of concentrated aqueous solutions is archetypal of this notion of activity. When the solution is diluted, the solutes dissolved in water are separated from each other by large quantities of water molecules, whose properties are little affected by these few extrinsic elements. Consequently, the activity of water in the aqueous solution is negligibly different from 1, and the activities of the dissolved ions are almost identical to their concentration: the activity coefficients are a little higher or lower than 1. In this case, the mutual influence of the solutes between them is perceived in an undifferentiated way, so that a single parameter, the ionic strength denoted I, is sufficient to describe most of the associated energy disturbances. It describes the half-sum of the electric charges distributed in a rather homogeneous way throughout the solution, which is itself electrically neutral. Several empirical models, based on interaction microphysics, calculate the activity coefficients across various concentration ranges, such as the Debye–Hückel ($I \leq 0.02$) or Davies ($I \leq 0.7$) models.

At high concentrations, the distance between the dissolved solutes decreases so that the amounts of "in-between" water are small, and their mutual influences become specific. The models integrate an increasingly large number of specific interaction parameters. The most successful model is that of Pitzer, which can consider binary (two ions of opposite charge and the same charge), or even ternary (three ions, with different variants) interaction parameters for ionic strengths beyond 3.5 mol/kg. The water activity deviates significantly from 1 (see section 1.2.2.2), and the ion activity coefficients change significantly. In addition, the formation of high salinity ion pairs further complicates the situation. The main problem with the Pitzer model is having the necessary data as the interactions become numerous and increasingly specific. Nevertheless, it offers a remarkable self-consistency so that its use could be extended for systems reaching ionic strengths of 30 or 40 mol/kg (Lach et al. 2015; Lassin et al. 2015).

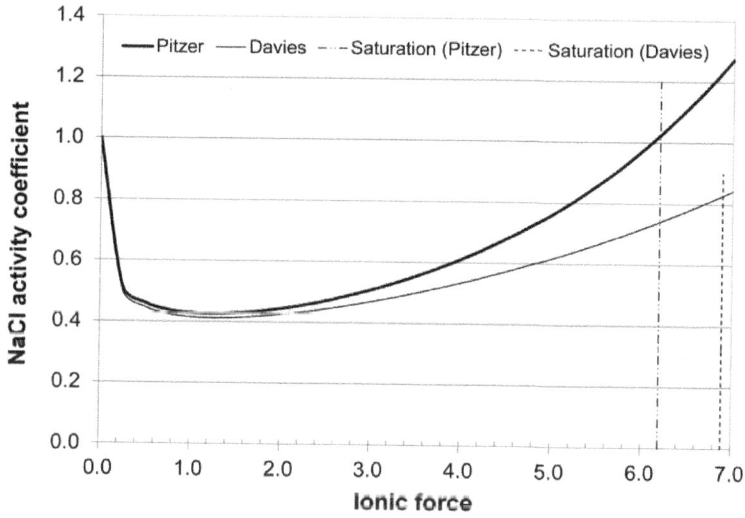

Figure 1.1. *Variation in the activity coefficient of the NaCl electrolyte according to two different models. Note that the saturation with respect to the halite depends on the model used*

For the calculation methods, the reader can refer to Steiger (2005) or Lassin et al. (2015) for an in-depth discussion of the Pitzer model, its evolution over time and modern developments. It is especially interesting to note that the activity coefficients of ions are generally less than 1 until saturation, which means that the solubility of the salts that dissociate into ions (halite, calcite, etc.) increases with the ionic strength. The presence of a minimum in the curve, which is also very typical, indicates that this effect varies greatly. In the case of NaCl, it is curious and usual to point out that at saturation, the activity coefficient is very close to 1. Another point of interest, the saturation with respect to the salt will be verified for varying ionic strengths depending on the activity coefficient model used. Here, this shows that Pitzer's model is the only one that accounts for the observations (saturation at 65 g/L), but also that a model that lowers the activity coefficients increases the solubilities. This very simple comparison illustrates the importance of these corrections when working with soluble salts.

Steiger (2005) correctly points out that extrapolating the Pitzer model beyond the saturation of the corresponding solid is not wise, because the terms involved are not conservative. He therefore proposes adjustments to explore the supersaturation of some salts.

1.2.2.2. *Water activity*

Water activity changes with the concentration and nature of dissolved species because they alter the dynamics of the intermolecular network that controls the thermodynamic properties of water. Therefore, the activity of water is a function of all the activities of the solutes, according to the following classical expression:

$$\ln a_{water} = -\phi \frac{M_{water}}{1{,}000} \nu\, m$$

where M_{water} is the molar mass of water, ϕ is the osmotic coefficient of the solute whose molality is m and which contains ν anions per mole of solute. In solutions whose ionic strength is less than 1, the water activity of any electrolyte can be estimated using that of a NaCl solution of the same concentration. For an ionic strength greater than 3, the Pitzer model in all its complexity must be used.

Besides the calculation method, the addition of solutes systematically leads to a decrease in activity with the increase in ionic strength. This indicates that water is less thermodynamically active, causing colligative properties (decrease in melting point, increase in boiling point, decrease in vapor pressure). The direct measurement of the variation in activity can be carried out by monitoring the drop in saturated vapor pressure of the solution with respect to the vapor pressure of pure water at the same temperature.

The dilution of a solvent by a solute is not the only way to vary the vapor pressure of water: a variation in the internal pressure of pure water also varies its saturated vapor pressure. This makes it possible to define the concept of osmotic pressure: it is the pressure of the free solvent (outside the water of solvation), which makes it possible to obtain the observed vapor pressure. As the activity drops, the vapor pressure is lower than that of pure water: the osmotic pressure of water in an aqueous solution is generally lower than the pressure of pure water.

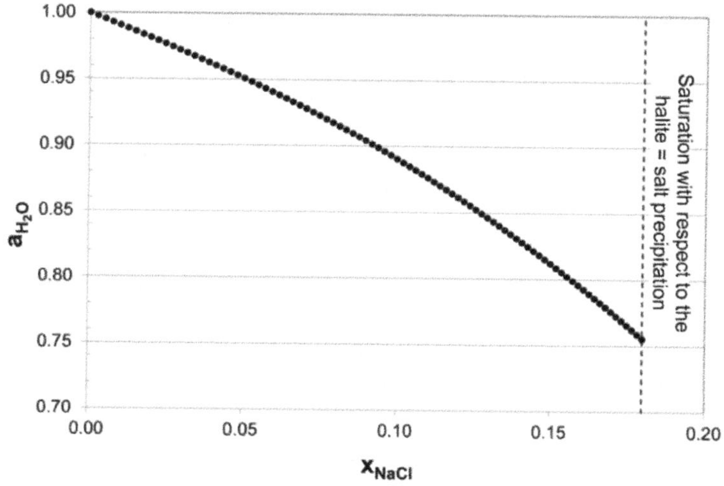

Figure 1.2. *Variation in water activity with NaCl molality until saturation with respect to the halite*

1.2.2.3. *Reaction stoichiometry*

Besides the activity of each species, it is necessary to emphasize the importance of the rule of defined proportions, which makes it possible to establish the conservation of mass during a chemical transformation (closed system):

$$\text{aCation}^+ + \text{bAnion}^- + \text{cH}_2\text{O} = \text{Cat}_a\text{An}_b.\text{cH}_2\text{O} \quad [1.3]$$

Written like this, the system is assumed to move from the initial state on the left to the final state on the right. This can be verified by calculating the free enthalpy of this transformation, of course integrating the proportions of each participant:

$$\Delta_R G_{T,P} = \overset{\circ}{\mu}_{Solid} + RT.\ln(a_{solid}) - \left[a.\overset{\circ}{\mu}_{Cation} + RT.\ln(\gamma_{cation}.C_{cation})^a + b.\overset{\circ}{\mu}_{anion} + RT.\ln(\gamma_{anion}.C_{anion})^b + c.\overset{\circ}{\mu}_{pure\ water} + RT.\ln(\gamma_{water}.x_{water})^c\right] \quad [1.4]$$

Standard state chemical potentials are standard state molar free enthalpies, and their values are usually tabulated. Moreover, in most cases, the activity of solids is equal to 1 (pure solid).

Equation [1.4] easily rearranges into:

$$\Delta_R G_{T,P} = \Delta_R G^{\circ}_{T,P} + RT.\ln\left(\frac{1}{(\gamma_{cation}.C_{cation})^{a}.(\gamma_{anion}.C_{anion})^{b}.(\gamma_{water}.x_{water})^{c}}\right) \quad [1.5]$$

The right-hand side of [1.5] highlights the ratio of the activities of the compounds that participate in the transformation, each to the exponent of its stoichiometric coefficient. This ratio is called reaction quotient, denoted as Q.

Note the importance of the stoichiometric term, especially when the salt considered involves ions of different valence, or for hydrates, since the number of water molecules is the exponent of the activity. Of course, when the ionic strength increases, the drop in water activity reduces the impact of this parameter.

1.2.2.4. *Temperature*

The temperature appears explicitly in expression [1.5] associated with the reaction quotient Q but is also less visibly involved in the calculation of the free enthalpy in the standard state. For each active species in the solution, the temperature where the transformation takes place must be considered for the calculation of its standard free enthalpy through the following expression:

$$\Delta G^{\circ}_{T,1bar} = \Delta G^{\circ}_{25°C,1bar} - S^{\circ}_{25°C,1bar} + \int_{298}^{T} Cp.dT - T.\int_{298}^{T} \frac{Cp}{T}.dT \quad [1.6]$$

where Cp is the heat capacity of the considered species. Data are widely available in the literature (Robie et al. 1978; Woods and Garrels 1987) for many minerals and soluble salts, making it possible to calculate these corrections, which are moderate for the majority of the ranges of interest (–20°C to +50°C).

1.2.2.5. *Pressure*

Sometimes the pressure is also a parameter of evolution of the system, and initially, it is expressed and quantified like the temperature in the calculation of the standard free enthalpy. The expression for a given compound is written as:

$$\Delta G^{\circ}_{T,P} = \Delta G^{\circ}_{T,1bar} + \int_{1}^{P} V^{\circ}.dP \quad [1.7]$$

Note that for all incompressible phases over the considered pressure range, the term on the right simplifies to a product of the volume times the pressure difference: $V^{\circ}\cdot(P_{final} - P_{initial})$.

Within a porous system, the pressure of each cohabiting phase can vary, which is typically expressed by a curved separation surface. Two examples can be easily given, with curved liquid–air interfaces and capillarity, and curved solid–liquid interfaces compared to the natural external forms of minerals.

1.2.2.5.1. Curvature of the water–air interface

In such a situation, liquid and air are at different pressures, the concavity of the curvature sinking toward the side of the lowest pressure phase. With the liquid being purely elastic, the deformation of the surface is a function only of this elasticity, which refers to the surface tension of the liquid γ (same notation as the activity coefficients, unfortunately). The mechanical equilibrium between the two phases at different pressures is reached when the Young–Laplace equality is verified:

$$\Delta P = P_{air} - P_{liquid} = \gamma_{liquid-air}\cdot\left(\frac{1}{R_1} + \frac{1}{R_2}\right) = \frac{2\gamma}{r_{spherical\,meniscus}} \quad [1.8]$$

where γ is the surface tension, so it should be specified which phases the interface separates. Strictly speaking, a water–water vapor interface has a different surface tension than a water–atmosphere meniscus. R_1 and R_2 are the two curvatures of the separation surface, which are equal for a spherical separation surface.

The Kelvin equation, the second law of capillarity, establishes the condition of chemical equilibrium between curved water and air at constant pressure:

$$ln\left(\frac{p}{p_s}\right) = \left[\frac{1}{RT}\int_1^P V_{water}\,.\,dP\right|_{capillary\;effect} + [\ln a_{water}]_{osmotic\;effect} \quad [1.9]$$

With the term dP being negative in the case of a concave interface toward the air, the vapor pressure of the capillary water decreases, and is added to that due to the dilution of the solvent by the solutes (see section 1.2.2.2). In terms of calculation, and to avoid counting the same effect several times, Lassin et al. (2005) recommend separating the two effects, considering the osmotic effect of salinity thanks to the water activity, while the capillary effect is integrated in the free enthalpy in the standard state. The term on the right of equation [1.9] brings out this distinction.

When a porous system saturated with water begins to dry out, the evaporation of the water and the associated loss of mass lead to the appearance of capillary menisci in the pores, the average radius of which verifies relation [1.8]. These menisci reflect the fact that the internal pressure of the water is lower than atmospheric pressure. This pressure contrast, this anisobarism of the water–air system, can be significant, and induce a capillary tension or negative pressure ($r < 1.45$ µm to 21°C). In particular, the chemical potential of the solution, that is, of the solvent and the solutes, is modified in proportion to this variation in internal pressure. The assumption is made that an aqueous solution constitutes a dielectric continuum so that the internal pressure in the solution has a continuous value. The pressures of water and solutes are equal, controlled by the Young–Laplace ratio (Mercury and Tardy 1997a, 1997b, 2001; Mercury et al. 2003).

As a result, it becomes easy to express the variation of free enthalpy due to the capillary effect:

$$\text{Gas} + \text{water} \leftrightarrows \text{Solute(s)} + \text{water} \qquad [1.10]$$

$$d\Delta_R G_{curvature} = \int_1^P V_{solute(s)} dP \qquad [1.11]$$

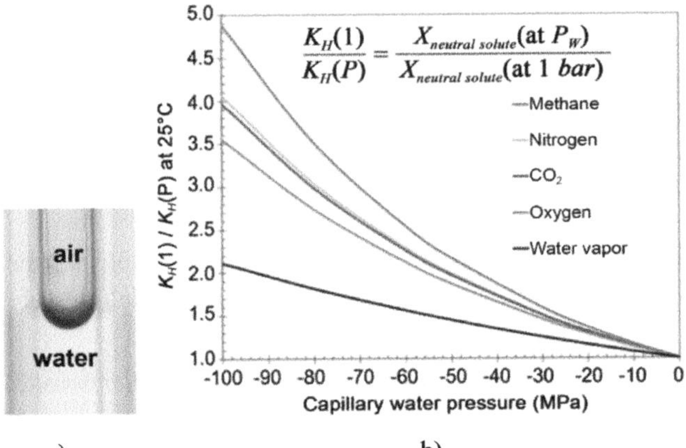

a) b)

Figure 1.3. *(a) Photograph of a cylindrical tube (∅ = 150 µm) and a capillary interface, concave toward the air. (b) Solubility of different gases in a solution under increasing capillary pressure. For a color version of this figure, see www.iste.co.uk/derluyn/salt.zip*

Using this formalism, we can calculate the variation in solubility of air gases in water as it becomes more capillary (Figure 1.3) (Mercury et al. 2004). A decrease in solubility is systematically predicted, via the salting-out effect, proportional to the molar mass of the gas. This thermodynamic prediction was recently confirmed by three series of independent experimental measurements (Lassin et al. 2016; Lidon et al. 2018; Hulin and Mercury 2019b).

In agreement with section 1.2.2.1, the variation in pressure of the solution imposes a correction on the activity coefficients. Lassin et al. (2005) showed that this correction increased with the ionic strength, using the model known as $\dot{B}$ (analogous to the Truesdell–Jones approach). Unfortunately, an explicit calculation at high ionic strength requires using the Pitzer model, whose calculation parameters as a function of pressure are not yet available.

Another point of interest is that the pressure value is determined by the fineness of the pore where the meniscus is placed and not according to the average radius in the porous system. Within a heterogeneous system, a meniscus trapped in a nanometric pore can put a large volume of solution under high capillary tension, because it is contained in much larger pores. (Mercury et al. 2003; Bouzid et al. 2011b; Hulin and Mercury 2019a, 2019b): it is the ink bottle effect (Figure 1.4) (Hillel 1998, p. 160).

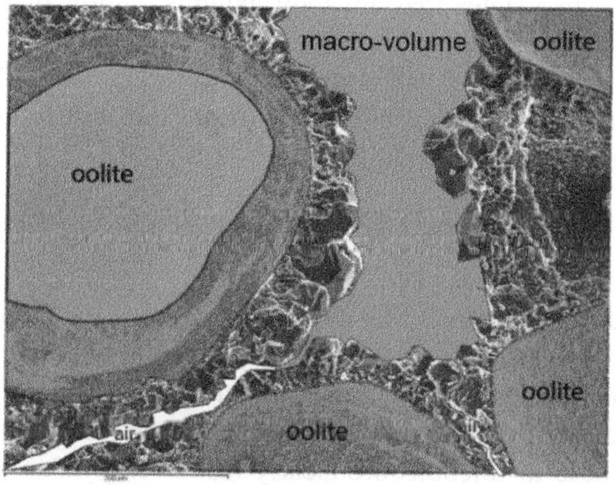

Figure 1.4. *Capillary trapping of a large amount of limited water by nanometric menisci in Young–Laplace and Kelvin equilibria, in a sedimentary rock (according to Hulin and Mercury (2019a)). For a color version of this figure, see www.iste.co.uk/derluyn/salt.zip*

1.2.2.5.2. Curvature of solid–solution interface

It is easy to write an equation of the type [1.8] for the curvature between the solid phase and the solution. We can cite the case of the Gibbs–Thompson equation, which links the variation in the melting temperature of a solid to its curvature. The

Ostwald–Freundlich equation is also widely used; it accounts for the excess solubility of small crystals compared to large crystals:

$$ln\frac{S}{S^{\circ}} = \frac{V_{solid}}{RT}.k.\frac{\gamma_{solid-solution}}{r_{solid}} \qquad [1.12]$$

where S and S° represent the solubility for a solid of finite and infinite dimensions respectively, and k is a form factor. It is analogous to the Kelvin equation [1.9] and the Gibbs–Thompson equation.

In a porous medium, crystallization occurs in a finite size medium, that of the pore. If the pore is small enough, [1.12] puts a very strong constraint on the size accessible to any crystal it contains: the solubility of the mineral formed will therefore depend on the size of the host pore. Rijniers et al. (2005) produced very nice illustrations of this reality and showed that the solubility of different salts varied significantly for pores smaller than 30 nm. The experiments were carried out at water saturation, which excluded any role of the capillary pressure of the water.

Equation [1.12] also states that anybody exhibits a form of equilibrium when growth occurs without a space constraint or material limitation. This shape is defined by its surface tension γ and corresponds to a reference solubility. It can be easily obtained for any polyhedral mineral by using the Wulff construction, which obeys the following mechanical equilibrium:

$$P_{solid} - P_{solution} = 2.\frac{\gamma_n}{h_n} \qquad [1.13]$$

where n is the index of any face of a Wulff crystal and h is the distance from the center of gravity to the outer surface of the face. Note that the term γ encompasses both the formation energy and the elastic contribution. The pressure increase in [1.13] therefore results from a chemical potential difference ($\Delta P = \Delta\mu/V$) and not surface tension. These distinctions are clearly discussed by Sanfeld and Steinchen (2000).

Consider a crystal growing in a medium under pressure $P_{solution}$. If one portion of the crystal grows faster than another, the radius of curvature there decreases. If γ is positive (Gibbs considers the case of negative surface tension; Gibbs 1928, p. 329), the pressure of the solid increases locally, which causes the material to flow back towards the interior of the crystal, the chemical potential increases, the growth slows down and a flat interface is obtained. This is what the Wulff construction (Figure 1.5) causes in the case of a polyhedral solid, where the surface tension is different from one face to the other. If the surface tension is constant, the mineral is spherical (spherule).

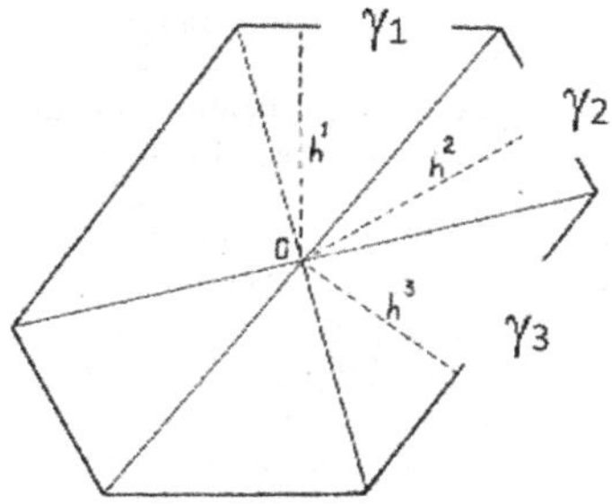

Figure 1.5. *Diagram of a crystal with the magnitudes of the Wulff construction (according to Defay and Prigogine (1951, p. 234))*

In practice, the chemical potential of solids is generally tabulated at 25°C and 1 bar: this is the chemical potential of the "natural" crystalline form under these conditions (*T*,*P*) of the initial solution. Any modification from this form of equilibrium reflects a change in chemical potential correlated with a change in pressure of the solid. Bouzid et al. (2011a) present an illustrative experiment with the coexistence of cubic and round NaCl, in connection with variable formation conditions.

We see by this that any modification with respect to the reference form is indicative of a thermochemical change when the observation is not taken in a too transient situation. Caution is required because the morphologies can also testify to kinetic factors. The main parameter is the diffusion of species at the interface. Typically, dendritic-like shapes that exhibit strong curvatures of solid–solution interfaces can be interpreted as resulting from high pressure gradients. However, they are also the consequence of mass deficit in certain directions of growth, when the growth rate at a tip is higher than the rate of diffusion in the solution. This term is also not purely kinetic, since the tip has a local excess of surface energy, which promotes precipitation. In the same way, the thermal effects linked to the dissipation of crystallization enthalpies can promote dendritic morphologies. Hollow crystals, called "hopper" crystals (Desarnaud et al. 2014), are also morphologies promoted by preferential growth rates, combining severe diffusional limitations with energetic factors like the presence of dislocations.

1.2.3. *Expression of the driving force*

The effects of temperature, pressure and composition can thus be dealt with in a thermodynamic way by evaluating the share of energy that each parameter brings or withdraws to the tendency of an aqueous solution to crystallize salts, and to therefore move beyond equilibrium. To consider the anisobaric effect linked to interface curvatures, it is possible to calculate the free enthalpy of each species under its own conditions, before combining them to evaluate the thermodynamic reaction potential. It should be noted that to know the directionality of the transformation, the free enthalpy is the potential to minimize, because if the system is anisobaric, the pressure relations express equilibrium conditions and they do not vary during the transformation.

To be in equilibrium ($\Delta_R G = 0$, no creation of entropy), it is necessary that the product of the activities, that is, the reaction quotient Q, is equal to $\Delta_R G°$. This is the $Q_{equilibrium}$ and this equilibrium reaction product is denoted K:

$$\Delta_R G^{\circ}_{T,P} = -RT.lnQ_{equilibrium} = -RTlnK \qquad [1.14]$$

Whenever the composition of the solution does not allow it to be at equilibrium, the actual reaction quotient can be compared to the reaction quotient required for it to be at equilibrium:

$$\Delta_R G_{T,P} = -RT.lnQ_{equilibrium} + RTlnQ_{actual} = RTln\frac{Q}{K} \qquad [1.15]$$

The ratio Q/K is the focal point of this whole demonstration. It helps to understand how the composition and the physicochemical "quality" (via the activity coefficients γ) control the sense in which the reaction is spontaneous. If $Q > K$, the thermodynamic potential is negative (creation of entropy) if one moves from right to left, for a reaction such as [1.3], a solid dissolves in such a solution. Under the same conditions, if $Q < K$, spontaneity is established from left to right, and it is the salt that crystallizes.

It is very easy to see that the more the ratio of the concentrations increases, the more the term Q is supposed to grow. However, the evolution of the ratio also depends on the value taken by the activity coefficients. If γ decreases, the proportion of this solute, which is active in the transformation, decreases. On the contrary, when γ increases, the solute is more thermodynamically active, that is, the interactions in solution increase its reactivity at constant composition. The temperature is considered the same throughout the whole system but affects each species according to their own sensitivity. The pressure can vary through interfaces characteristic of interfacial equilibria, as indicated above. To illustrate this effect, and certainly the least frequently studied, we can compare the behavior of mirabilite and thenardite, the decahydrate and anhydrous forms of sodium sulfate: the hydrated form is much more strongly impacted than the anhydrous form by the variation in pressure of the solution (Figure 1.6(a)). We can also look at the change in solubility of NaCl, if its external form is independent (anisobaric case) or dependent (isobaric case) of the pressure of the liquid (Figure 1.6(b)). Bouzid et al. (2011a) presented an experiment confirming that halite bears a different shape depending on the formation conditions.

The degrees of freedom of a system not saturated with water, with capillary interfaces, are greater than those of a saturated system. Water–air menisci give occluded water supersolubility with respect to gases and solids. Capillary water can also trigger precipitation of nominally undersaturated solids, provided they change their external shape. These effects are greater when the mineral is hydrated. We can thus see that the external forms and the degrees of hydration are good markers of the formation conditions.

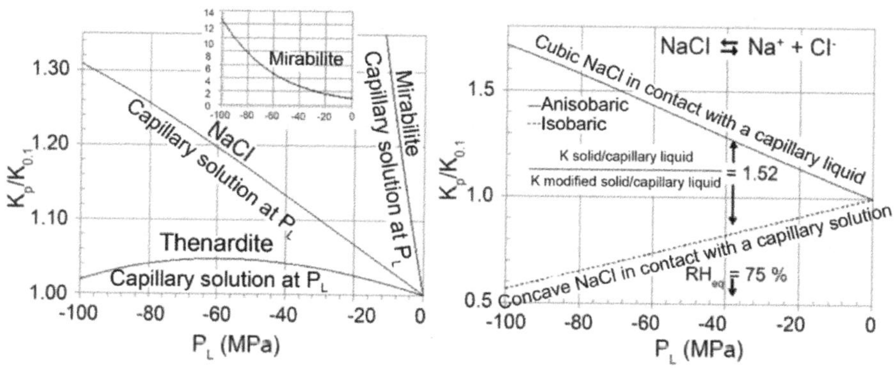

a) b)

Figure 1.6. *(a) Effect of solid (ambient pressure) – solution (capillary pressure) anisobarism for three different solids (according to Hulin and Mercury (2019a)). (b) Solubility of the halite in isobaric and anisobaric situations (according to Bouzid et al. (2011a))*

1.3. Stability, metastability, and instability

Thermodynamics can predict the state of equilibrium by calculating the minimum value of the thermodynamic potential that the system can reach by creating entropy. This assumes that nothing prevents this transformation from taking place, once the created entropy is made available. However, in many cases, the transformation must expend additional energy to start the transformation, which is called the activation energy E_a. To illustrate this point in the situation that interests us, the first crystal that is made is generally small, and therefore has a higher surface/volume ratio (S/V) than a big crystal. However, the surfaces of a solid have an additional energy term, called surface energy, which is added to the energy density of the compound. In this sense, the higher the *S/V* ratio, the more energy the production of the corresponding crystal requires. Some additional energy must therefore be given to a solution to make the first crystal and all its interfaces. This is typically the situation described by [1.12], which greatly exceeds the case of intrapore growth in fine pores.

Whether it is a problem of the existing solid surface or not, we immediately see that the presence of pre-existing pore walls or particles in suspension drastically changes the situation with respect to this triggering element. The problem becomes a little more subtle with the energy compatibility between the crystal structure of the existing surface compared to that of the crystal to nucleate and grow. For this, the surface tension is the relevant parameter. We can thus define a homogeneous nucleation, which only depends on the interfacial energy of the nucleus, the supersaturation (i.e. the frequency of "favorable collisions") and the temperature. If we consider a heterogeneous nucleation, the interfacial energy is expressed differently to consider the role of the mediating substrate (Stumm 1992):

homogeneous case: $\Delta G_{SURF} = \gamma_{NL}.A$

heterogeneous case: $\Delta G_{SURF} = \gamma_{NL}.A_{NL} + (\gamma_{NS} - \gamma_{SL}).A_{NS}$

where N is the nucleus, L is the liquid, S is the substrate and A is the surface area.

Heterogeneous nucleation is generally accompanied by a lowering of the energy barrier because, most often, $\gamma_{NS} < \gamma_{SL}$. The borderline case is epitaxial growth, when the host surface of the substrate corresponds very well, even perfectly, to the crystal to be formed ($\gamma_{NS} \rightarrow 0$ and $\gamma_{SL} \rightarrow \gamma_{NL}$).

The characteristic equation usually written is that of the classical nucleation theory, which gives the rate of nucleation J in a volume for 1 mole:

$$J = J_0 . exp \frac{-W*}{RT} = J_0 . \exp \frac{-k.\gamma_{NL}^3.V^2}{R^3T^3\left(ln^{Q}/_{K}\right)^2} \qquad [1.16]$$

with J_0 is the pre-exponential factor, $W*$ is the activation energy of a spherical germ with a finite interface V is the molar volume of the nucleus (the other factors have already been mentioned).

These developments provide a good understanding of the thermodynamic reasoning for metastable states. The higher the energy barrier, the longer the activation time, that is, the time required for a fluctuation to provide sufficient energy to overcome it. This time is a function of the kinetics of formation of germs in the system, which is the product of a frequency factor and the density of matter. When the activation energy is surpassed, the system becomes unstable, and nothing keeps it from converging to the equilibrium point. Once nucleation has been activated, the initial solution sees many germs appear, then each begins to grow according to the laws of crystal growth, which obey the principle of energy minimization (see Wulff crystal). Depending on the amount of mass available locally and the growth rate relative to the diffusion rate of the species in the solution, the external forms will be automorphic or not.

In our case, the energy barrier is provided by the critical chemical supersaturation that triggers crystallization. When the excess chemical energy, associated with fluctuations, usually of the thermal type, becomes commensurate with the energy barrier at any point in the system, nucleation is activated, and the phase transition becomes spontaneous and takes place instantaneously: it can no longer be stopped until it has reached equilibrium concentration. Equation [1.16] indicates that the greater the interfacial tension of a solid and its volume, the higher the critical supersaturation will be. Note that this purely thermodynamic term does not consider the rate at which the supersaturation changes. However, it is a parameter that is well known experimentally, with the supercooling rate for example. The rate of change generates supersaturation gradients in the solution, but also compositional gradients around the seeds, due to diffusion. Crystallization is a process that must be studied locally.

A final aspect that can be reported is the occurrence of metastable phases that form first, before the stable phase: disordered dolomite before ordered dolomite (Navrotsky 1999), amorphous calcium carbonate, then vaterite or aragonite before calcite (Lassin et al. 2018), heptahydrate before sodium sulfate decahydrate (Steiger and Asmussen 2008), etc. In fact, the phase that forms first is the one that nucleates the most easily and not the one that is the most stable (Navrotsky 2004). Kinetic and thermodynamic arguments support this "rule".

1.4. Kinetics of phase transformations

Thermodynamics is a science of initial and final states but does not consider the path taken to get from one to the other. Therefore, it is an effective science to predict the transformation that should occur, but it cannot say if the transformation occurs, let alone when, even taking into account the activation energy. Chemical kinetics thus usefully completes this.

1.4.1. *Characteristic equation*

The rate of the phase transition is a function of the supersaturation reached and depends on the diffusion of the species which "feed" this supersaturation and on the total volume of solution, but also on other more strictly kinetic terms. It is usual to write a global kinetic law (Palandri and Kharaka 2004) that groups the active parameters over the time required for a chemical system to reach equilibrium:

$$R = k_0 . A' \, exp\left(-\frac{E_a}{RT}\right) . \frac{A^{\circ}}{V}\left(\frac{m}{m^{\circ}}\right)^{n} . \prod_i a_i^{n_i} . \left(1-\left(\frac{Q}{K}\right)^{e_1}\right)^{e_2} \qquad [1.17]$$

where k_0 is the intrinsic kinetic constant in mol/(sec.m^2) ; A' is a pre-exponential factor and E_a is the activation activation; $A°$ is the surface area initially offered for transformation; V is the volume of the active solution; $m/m°$ is the lost mass correction; n, n_i, e_1, e_2 are empirical exponents; Πa_i encompasses all the possible effects related to the composition.

The intrinsic constant is the reaction time per unit area. Measured in the laboratory under conditions far from equilibrium, it gives the upper rate limit. The Arrhenius term that follows corrects this rate for the effect of temperature. A' and E_a are generally considered independent of temperature and are measured experimentally. The known values of activation energies for typical solids vary from a few tens to a few hundreds of kJ/mol. An $E_a \leq 20$ kJ/mol means that the reaction time will be rather limited by the diffusion of the transformation partner species. An $E_a \geq 80$ kJ/mol will indicate a need for bond breaking in the transformation, which will limit reaction time.

The next parameter $A°/V$ is crucial when studying a dissolution, and the notion of reactive surface is still much debated and under study. In the case of crystallization, its role is much less important except in the case of epitaxial growth (from a substrate identical to the nucleus, homoepitaxy, or different, heteroepitaxy). On the other hand, the mass ratio at time t of the initial mass makes it possible to correct the reactive surface area of the mass lost during a dissolution.

The activity product of the ions indicated in [1.17] makes it possible to integrate the role of inhibitors or catalysts, or even the role of pH, on the reaction rate. Obviously, kinetics deals with transformation mechanisms so that the speed of movement of these species with respect to the reaction site where they occur (formation of the activated complex, sorption, etc.) is also a parameter that will contribute to the weight of this parameter.

The last term expresses the deviation from thermodynamic equilibrium with respect to the reaction studied, weighted by empirical exponents for reasons that will not be detailed here. Note that this is a local *Q/K* ratio, at the very place where the reaction advances, lowering itself by consuming the supersaturation. This is why it is important to measure the local concentration in the domains where the phase transition takes place, and to be able to characterize the diffusive and/or advective fluxes, which move the dissolved mass toward/from this local domain.

1.4.2. *Chemistry-transport coupling*

This chapter has so far focused on a closed system, whose composition changes with evaporation (for example) and whose (over)saturation increases with water loss. However, real systems are of course open, combining phenomena of advection, diffusion and dispersion. Advection occurs even in the unsaturated zone, under the effect of capillary pressure gradients, which bring mass back to the most strongly bound water–air menisci. These are effects that play a significant role in CO_2 storage and can contribute to decreasing the injectivity in the nearby carbon sink (André et al. 2014; Ott et al. 2014). Fickian diffusion combined with tortuosity diffusion is the first barrier to crystal growth. Note that in the case of concentrated solutions, the Fickian diffusion lacks generality because it is strictly applicable only in an infinitely dilute solution with uncharged species. Electrochemical migration makes it possible to study the role of charged species, by including different diffusion coefficients for each species while preserving the electroneutrality of the solution (Steefel 2008). However, this effect has been little studied and is very rarely considered. The second limit of the Fickian approach is encountered when working with concentrated solutions, which is the case with soluble salts. The driving force of the movement is no longer the concentration gradient alone, but the chemical potential gradient, which brings up terms of activity coefficients. As a result, the concentration term, which increases near the interfaces, can be reduced or even reversed by the activity coefficient term, so that counter-concentration diffusions can be considered (Steefel 2008).

The chemistry-transport coupling integrating the kinetic aspects could be the subject of a specific chapter. The use of dimensionless terms is particularly useful for looking at the relative importance of dynamic processes and accounting for reaction fronts. These studies also make it possible to specify the exact regime of the crystallization studied, and its exact control regarding the circumstances: control by transport or reaction, or even mixed control. For a complex process, the limiting step is obviously the slowest one. In fact, without coupling, it becomes very difficult to qualify the time constraints, usable mass and impacted space. It is also impossible to scale reaction rates in a heterogeneous porous medium by ignoring these couplings. Finally, dissipative structures (reactive fingers, Liesegang rings and oscillating reactions, chemomechanical replacements, etc.) most often result from initial conditions far away from equilibrium with very large driving forces and available masses. They cannot be understood without an approach coupling reaction and transport, but also involving thermal and mechanical stress fields.

1.5. Conclusion

This brief overview aims to establish the conceptual landscape within which crystallization phenomena can be interpreted, understood and quantified within porous media. One of the main obstacles in this field is the consistency of two rather disjoint approaches, with fluid mechanics, dominated by a vectorial vision linked to force (intensity and direction), with respect to thermodynamics, dominated by a scalar and energetic vision, whose directionality is ensured by the maximum entropy (second law). The thermodynamics of irreversible processes, especially in its linear approximation, makes it possible to linearly link the flow to its driving gradient (typically the chemical potential) with phenomenological coefficients, measurable quantities characteristic of the coupling.

Inside the reactive compartment itself, the combination of thermodynamic and kinetic terms must consider subtle co-influences and feedbacks, which are not always very easy to separate. Typically, the surface tension of the solid, or interfacial tension, is the key parameter both in the reaction energetics, but also in the paths followed. It is a local term, which can therefore be discretized within a complex process, while also capable of being scaled up.

These proposed frameworks offer interesting benchmarks for the construction of coupled thermokinetic models, which make it possible to follow the composition of driving forces and flows, on scales of time and space favorable to this mechanistic approach. We cannot say that a unified theory is already defined to comprehend the discrete scale of the pore. Further work will be required to scale up and integrate these approaches on a continuous scale.

1.6. References

André, L., Peysson, Y., Azaroual, M. (2014). Well injectivity during CO_2 storage operations in deep saline aquifers. Part 2: Numerical simulations of drying, salt deposit mechanisms and role of capillary forces. *Int. J. Greenhouse Gas Control*, 22, 301–312.

Bouzid, M., Mercury, L., Lassin, A., Matray, J.-M., Azaroual, M. (2011a). In-pore tensile stress by drying-induced capillary bridges inside porous materials. *J. Colloid Interface Sci.*, 335, 494–502.

Bouzid, M., Mercury, L., Lassin, A., Matray, J.-M. (2011b). Salt precipitation and trapped liquid cavitation in micrometric capillary tubes. *J. Colloid Interface Sci.*, 360, 768–776.

Defay, R. and Prigogine, I. (1951). *Tension superficielle et adsorption*. Desoer, Liège.

Desarnaud, J., Derluyn, H., Carmeliet, J., Bonn, D., Shahidzadeh, N. (2014). Metastability limit for the nucleation of NaCl crystals in confinement. *J. Phys. Chem. Lett.*, 5(5), 890–895.

Gibbs, J.W. (1928). *Collected Works. Vol. I: Thermodynamics*. Longmans, Green & Co, London.

Hillel, D. (1998). *Environmental Soil Physics*. Academic Press, New York.

Hulin, C. and Mercury, L. (2019a). Capillarity-driven supersolubility in dual-porosity systems. *Geochim. Cosmochim. Acta*, 252, 144–158.

Hulin, C. and Mercury, L. (2019b). Regeneration of capillary water in unsaturated zones. *Geochim. Cosmochim. Acta*, 265, 279–291.

Lach, A., André, L., Lassin, A., Azaroual, M., Serin, J.-P., Cézac, P. (2015). A new Pitzer parameterization for the binary NaOH–H_2O and ternary NaOH–NaCl–H_2O and NaOH–LiOH–H_2O systems up to NaOH solid salt saturation, from 273.15 to 523.15 K and at saturated vapor pressure. *J. Solution Chem.*, 44, 1424–1451.

Lassin, A., Azaroual, M., Mercury, L. (2005). Geochemistry of unsaturated soil systems: Aqueous speciation and solubility of minerals and gases in capillary solutions. *Geochim. Cosmochim. Acta*, 69, 5187–5201.

Lassin, A., Christov, C., André, L., Azaroual, M. (2015). A thermodynamic model of aqueous electrolyte solution behavior and solid-liquid equilibrium in the Li-H-Na-K-Cl-OH-H_2O system to very high concentrations (40 molal) and from 0 to 250°C. *Am. J. Sci.*, 315, 204–256.

Lassin, A., Marty, N.C.M., Gailhanou, H., Henry, B., Trémosa, J., Lerouge, C., Madé, B., Altmann, S., Gaucher, E.C. (2016). Equilibrium partial pressure of CO_2 in Callovian-Oxfordian argillite as a function of relative humidity: Experiments and modelling. *Geochim. Cosmochim. Acta*, 186, 91–104.

Lassin, A., André, L., Devau, N., Lach, A., Beuvier, T., Gibaud, A., Gaboreau, S., Azaroual, M. (2018). Dynamics of calcium carbonate formation: Geochemical modeling of a two-step mechanism. *Geochim. Cosmochim. Acta*, 240, 236–254.

Lidon, P., Marker, S.C., Justin, W.J., Williams, R.M., Zipfel, W.R., Stroock, A.D. (2018). Enhanced oxygen solubility in metastable water under tension. *Langmuir*, 34, 12017–12024.

Mercury, L. and Tardy, Y. (1997a). Negative pressure and thermodynamic properties of capillary water. A concise review paper. *C.R. Acad. Sci.*, 324(11), 863–873.

Mercury, L. and Tardy, Y. (1997b). Physicochemical features of water in capillaries and fog water droplets. *C.R. Acad. Sci.*, 325(12), 947–954.

Mercury, L. and Tardy, Y. (2001). Negative pressure of stretched liquid water. Geochemistry of soil capillaries. *Geochim. Cosmochim. Acta*, 65, 3391–3408.

Mercury, L., Azaroual, M., Zeyen, H., Tardy, Y. (2003). Thermodynamic properties in metastable systems under negative or positive pressures. *Geochim. Cosmochim. Acta*, 67, 1769–1785.

Mercury, L., Pinti, D.L., Zeyen, H. (2004). The effect of the negative pressure of capillary water on atmospheric noble gas solubility in ground water and palaeotemperature reconstruction. *Earth & Planetary Sci. Lett.*, 223, 147–161.

Navrotsky, A. (1999). High temperature reaction calorimetry applied to metastable and nanophase materials. *J. Therm. Anal. Cal.*, 57, 653–658.

Navrotsky, A. (2004). Energetic clues to pathways to biomineralization: Precursors, clusters, and nanoparticles. *PNAS*, 101(33), 12096–12101.

Ott, H., Andrew, M., Snippe, J., Blunt, M.J. (2014). Microscale solute transport and precipitation in complex rock during drying. *Geophys. Res. Lett.*, 41, 8369–8376.

Palandri, J.L. and Kharaka, Y.K. (2004). A compilation of rate parameters of water-mineral interaction kinetics for application to geochemical modeling. Report, U.S. Geological Survey.

Rijniers, L.A., Huinink, H.P., Pel, L., Kopinga, K. (2005). Experimental evidence of crystallization pressure inside porous media. *Phys. Rev. Lett.*, 94, 075503.

Sanfeld, A. and Steinchen, A. (2000). Surface energy, stress, capillary-elastic pressure and chemical equilibrium constant in nanoparticles. *Surface Science*, 463, 157–173.

Steefel, C.I. (2008). Geochemical kinetics and transport. In *Kinetics of Water-rock Interactions*, Brantley, S.L., Kubicki, J.D., White, A.F. (eds). Springer Science, New York.

Steiger, M. (2005). Crystal growth in porous materials I: The crystallization pressure of large crystals. *J. Cryst. Growth*, 282, 455–469.

Steiger, M. and Asmussen, S. (2008). Crystallization of sulfate phases in porous materials: The phase diagram Na_2SO_4-H_2O and the generation of stress. *Geochim. Cosmochim. Acta*, 72, 4291–4306.

Stumm, W. (1992). *Chemistry of the Solid-water Interface*. Wiley, New York.

2

Crystallization Pressure

Noushine SHAHIDZADEH
IoP, Van der Waals-Zeeman Institute,
University of Amsterdam, The Netherlands

2.1. Introduction

A significant part of the alteration of historical monuments, bridges, roads or even more generally of buildings is caused by salts present in the environment and within these structures themselves. Some important examples include ancient structures such as the Valley of the Kings in Egypt (Nicolaescu 2009) and the famous monuments in Petra, Jordan, but also many contemporary constructions (Goudies 1997). Due to climatic variations (temperature, relative humidity [RH]), the crystallization process of salts dissolved in fluid saturating the porous space of the materials concerned (stones, bricks, concrete, etc.) causes considerable damage with not only aesthetic consequences, but also structural ones.

With the recent awareness of the consequences of climate change, inducing extreme variations in environmental conditions, the safeguarding of monuments has regained international interest: these variations are already affecting many structures and artifacts that are part of our cultural heritage and are likely to affect many more in coming years (Colette et al. 2007; Sabionni et al. 2008). Some risk factors associated with such a change include sea level rise, intense rain, flooding and changes in humidity cycles, greatly amplifying phase changes in freeze-thaw cycles and in the salt crystallization process. A particular example is that of old buildings. As these are made up of more porous materials than contemporary buildings, any increase in soil moisture due to temperature and/or sea level rise would result in greater mobilization of salt within the structures (Shahidzadeh-Bonn et al. 2010; Espinoza-Marzal 2013; Flatt et al. 2014; Desarnaud et al. 2015), which would increase the risk of damage to these buildings during periods of wetting/drying.

The salts present in built structures can have very diverse origins. Some building materials, such as lime mortar and limestone, naturally contain salts (e.g. the Narbonne cathedral in Southern France). Constructions located near the sea or agricultural land (presence of salts in chemical fertilizers) constitute a large part of the structures affected by these phenomena. A typical case is that of Venice, whose buildings are permanently partially submerged in lagoon salt water. The salt-laden water from the subsoil saturates the facades by capillary rise, leaving the salt to crystallize during the drying cycles. In addition, the salting of roads during the winter is one of the main sources of salt affecting buildings. For example, each winter, between 600,000 and 1,650,000 tons of salt are spread across France. De-icing salts cause damage to roads, contaminate soil, groundwater and surface water, and are absorbed by buildings located near the treated areas. The dissolved salts are then transported during the drying periods to evaporation zones (such as facades) where they crystallize.

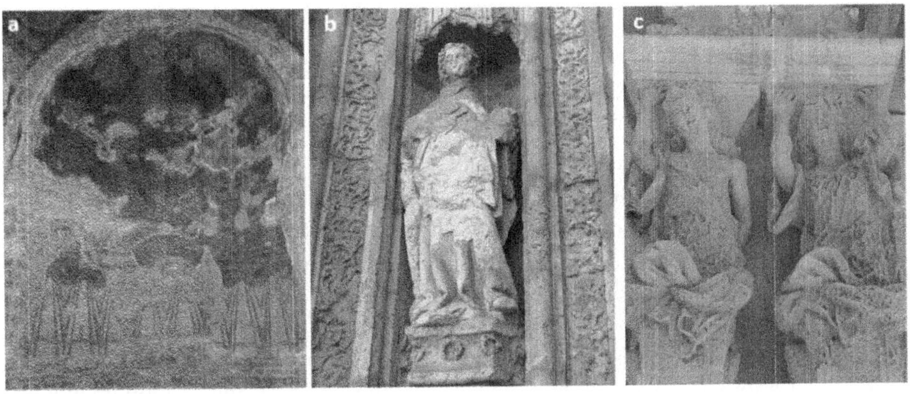

Figure 2.1. *Mechanical damage, leading to the destruction of materials due to subflorescence (crystallization within a porous network). Damage in the form of (a) peeling paint (Roman Forum, Italy), (b) granular disintegration (Cordoba, Spain) and (c) honeycomb weathering (Lecce, Italy) (pictures: Noushine Shahidzadeh). For a color version of this figure, see www.iste.co.uk/derluyn/salt.zip*

When crystallization occurs on the surface of a material, it is known as *efflorescence* (Veran Tissoire et al. 2012, 2014). The conditions for efflorescence formation will be discussed in more detail in Chapter 3 of this book. In contrast, subsurface crystallization in confinement (within the porous network) is called *subflorescence*. It can lead to mechanical damage including the destruction of materials (Rijniers et al. 2005; Lubelli et al. 2006; Shahidzadeh-Bonn et al. 2010; Derluyn et al. 2014; Flatt et al. 2014). This damage leads to the degradation of the surface condition of masonry by causing the flaking of paint on facades and frescoes, or even cracking, granular disintegration, or honeycomb weathering of buildings (Figure 2.1).

2.2. Crystallization pressure

The weathering of rocks due to salts, known as haloclasty, can be linked to three types of processes: crystal growth in confinement, thermal expansion and crystal hydration. Nevertheless, the main mechanism that explains this damage is related to the crystallization pressure. The first arguments, explaining that the growth of a crystal in confinement generates stresses, date back more than 150 years. The pioneering work of Lavalle (1853) and later that of Becker and Day (1905) report that a crystal, confined between two slides in its saturated solution, can lift the upper slide even if an external pressure is applied to it (Figure 2.2). To explain these observations, they then assumed – without providing experimental evidence – that the crystal could grow and overcome external forces only if the confined crystal surface remained in contact with a film of supersaturated salt solution (Taber 1916; Correns 1949; Scherer 1999; Flatt et al. 2007). If this hypothesis turns out to be correct, the question arises as to the origin of the force responsible for the stability of this film throughout the growth of the crystal.

This is why the existence of this constraint has been hotly debated for years. The work of Weyl (1959) and more recently of Desarnaud et al. (2013) report a different scenario to that of Becker. They suggest that it is the dissolution of the constrained crystal face that favors the growth of the unconfined faces. The argument is based on two points. On the one hand, a growing crystal brought into contact with the wall will stop growing in favor of the unconfined faces, which will continue to grow. On the other hand, if the confined face is constrained, the solubility in this region will be higher compared to the unconfined faces and therefore it should dissolve rather than expand (Figure 2.3).

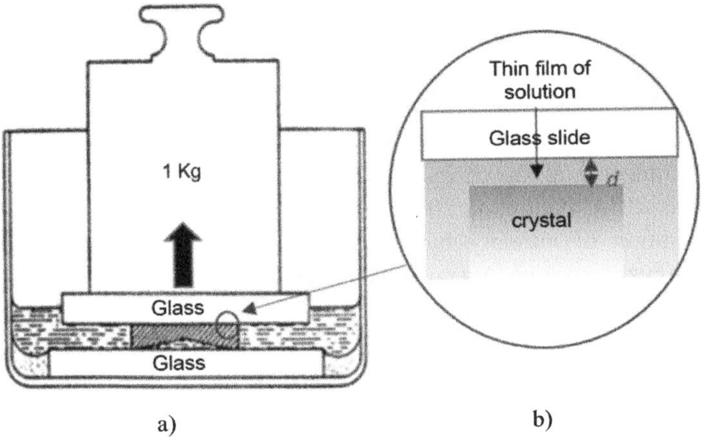

Figure 2.2. *(a) Diagram of Becker's experiment published in 1905, showing the growth of a crystal between two glass slides and the lifting of the upper slide loaded with a mass of 1 kg (Becker and Day 1905). (b) Becker's hypothesis schematized to justify this growth. For a color version of this figure, see www.iste.co.uk/derluyn/salt.zip*

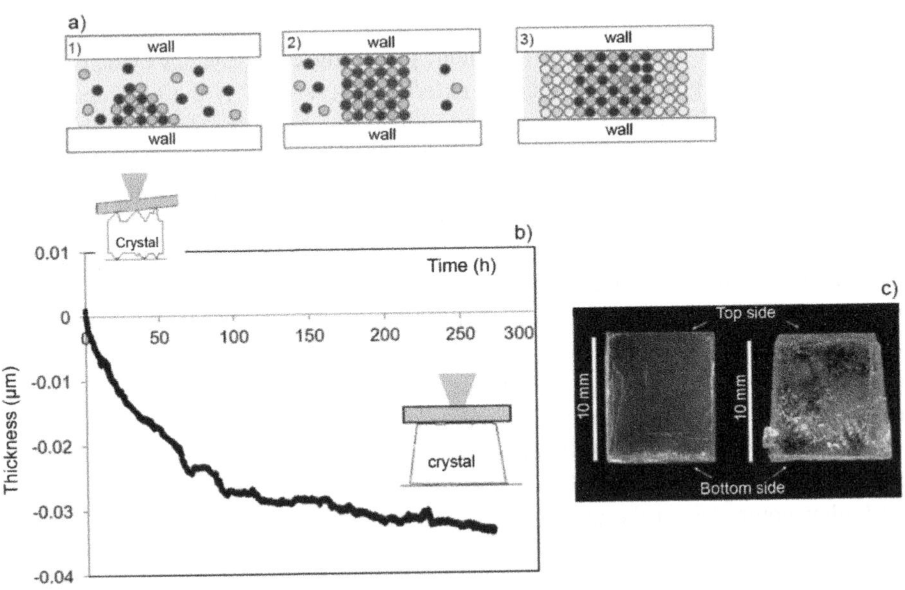

Figure 2.3. *Growth and dissolution of a crystal under stress. For a color version of this figure, see www.iste.co.uk/derluyn/salt.zip*

COMMENTS ON FIGURE 2.3.– *(a) Illustration of the growth of crystal faces in confinement before and after contact with the confining surfaces. (b and c) The experiment by Desarnaud et al. (2013) going against Becker's theory. The graph in (b) shows the experimental results of the dissolution of the confined face of the KCl crystal and (c) shows the trapezium-shaped growth of the unconfined faces in contact with the saturated solution during evaporation (Desarnaud et al. 2013).*

The first thermodynamic equation for crystallization pressure, in its simplest form, was expressed (Correns 1949) by the chemical potential difference between the salt molecules in the crystal and in the supersaturated salt solution. In any crystallization process, the reason for the transfer of ions from solution to the crystal phase is that the free energy of the initial phase of the solution is greater than the sum of the free energies of the crystal phase and that of the final solution (Mullin 1992). Therefore, for the precipitation reaction:

$$\nu_A A^{z_A^+} + \nu_B B^{z_B^-} \Leftrightarrow A_{\nu A} B_{\nu B} \qquad [2.1]$$

The driving force can be expressed as the change in the chemical potential of the crystallizing species ($\Delta\mu = \mu_l - \mu_s$), which measures the loss or gain of free energy for molecules transitioning from one phase to another. For dissolved salt, the chemical potential is defined as:

$$\mu_l = \mu_l^0 + \nu RT\, ln\, a$$

$$a = a_A^{\nu A} a_B^{\nu B} \qquad [2.2]$$

where μ_l^0 is the chemical potential in the standard state, ν is the sum of anions and cations ($\nu_A + \nu_B$), T is the absolute temperature and R is the gas constant; here a is the activity of dissolved salt with a_A and a_B, respectively, the activities of the cation and the anion. Therefore, the change in chemical potential ($\Delta\mu$) is directly related to the activity of the product (Steiger 2005):

$$\Delta\mu = \mu_l - \mu_c = \nu RT\, ln\, ln\, a\, / a_0 \qquad [2.3]$$

where μ_l and μ_c are the chemical potentials of a molecule in solution and the crystalline phase, ν is the total number of ions per dissolved molecule (2 for NaCl and 3 for Na_2SO_4), R is the ideal gas constant (8.317 $J \cdot K^{-1} \cdot mol^{-1}$) and T is the temperature (K). In dilute solutions, the activities in solution are often calculated using the concentrations in solution, neglecting the effect of the ionic activity coefficient. Thus, in the Correns equation, the crystallization pressure exerted by a salt crystal is simply related to the degree of supersaturation of the solution in which it grows, knowing that $\Delta\mu = V_m \Delta P$ (Atkins 1992), where ΔP is the pressure exerted by the crystal growth, that is, the crystallization pressure (MPa), and V_m is the molar volume of crystallized salt ($cm^3 \cdot mol^{-1}$).

Subsequently, corrections were made to this expression in order to consider the non-ideal character of supersaturated solutions:

$$\Delta P = \frac{\nu RT}{V_m}\left(\ln\frac{m}{m_0} + \ln\frac{\gamma_\pm}{\gamma_{\pm,0}} + \frac{\nu_0}{\nu}\ln\frac{a_w}{a_{w0}}\right) \qquad [2.4]$$

Therefore, $\Delta P = P_C - P_0$, where P_C is the pressure required to suppress crystal growth and P_0 is the pressure of the liquid. The first term in parentheses m/m_0 describes supersaturation in terms of concentration (also called S), with m as the concentration of the solution in contact with the crystal under pressure (therefore supersaturated) ($mol \cdot cm^{-3}$) and m_0 as the concentration of the saturated solution at equilibrium ($mol \cdot cm^{-3}$). The second and third terms describe the non-ideal behavior of concentrated salt solutions. m_0, $\gamma_{\pm,0}$ and $a_{w,0}$ refer to the molality, mean activity coefficient and water activity of saturated solutions, respectively. ν_0 has the value of 0 for anhydrous solids and 7 or 10 as an example for sodium sulfateheptahydrate and decahydrate.

Directly measuring the pressure exerted by a growing crystal in confinement is very difficult and this led to questioning the very concept of the measurement. Recently, considerable effort has been made to better understand the evolution of this pressure in confinement (Hamilton 2010; Sekine 2011; Royne 2012; Desarnaud 2013; Desarnaud et al. 2016; Naillon et al. 2018; Dziadkowiec et al. 2019).

2.3. Direct measurement of the stress generated during crystallization in confinement

Research published in recent years has provided a better understanding of the concept of crystallization pressure and the conditions for its development (Hamilton 2010; Sekine 2011; Royne 2012; Desarnaud 2013; Desarnaud et al. 2016; Naillon et al. 2018; Dziadkowiec et al. 2019). In order to better understand the concept, we will describe here the experiments allowing the direct measurement of stress generated by crystallization in confinement (Desarnaud et al. 2016).

The experimental measurements are based on an original device, built in the laboratory, combining an inverted microscope and a force measuring device (rheometer). The evaporation of the saline solution confined between two slides, enables the visualization of primary nucleation and the growth of the crystal while directly measuring the force exerted by the crystal during its growth. By primary nucleation, we mean here the precipitation of a single crystal in a homogeneous salt solution without a pre-existing crystal (Figure 2.4). In this experiment, the supersaturation of the solution at the time of precipitation is not imposed and no load is applied to the upper face of the crystal; previous attempts to measure crystallization pressure have been made during secondary nucleation (growth of a pre-existing crystal in salt solution), for which very modest supersaturations would develop (Correns 1949; Desarnaud et al. 2013).

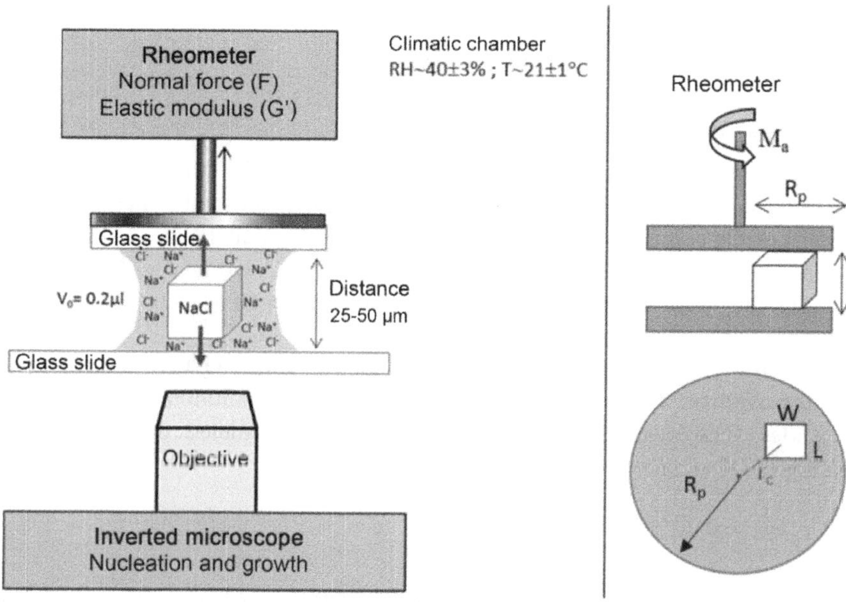

Figure 2.4. *Experimental setup allowing the measurement of the force generated during crystallization in confinement. For a color version of this figure, see www.iste.co.uk/derluyn/salt.zip*

The experimental setup illustrated in Figure 2.4 is placed in a climatic chamber to control the RH. The experiments are carried out for typical RH values of 40–55%. A very small drop ($V \approx 0.2$ µL) of saline solution (NaCl $m_s = 6.16$ molal) or KCl ($m_s = 4.56$ molal) close to saturation ($S = m/m_0 = 0.9$) is placed on a clean glass slide on an inverted microscope. The drop is confined by capillary action by approaching a second glass slide from above and confined between the two slides by capillary action. A charge-coupled device (CCD) camera connected to the microscope makes it possible to follow the evaporation of this drop, the precipitation and the growth of the crystal over time in the drop. The upper slide is attached to the rheometer allowing: (i) the control and measurement of the space h between the slides, (ii) measurement of the normal force during crystal growth, and (iii) estimation of the contact area of the crystal between the two slides. This latter estimate is made by applying very small oscillations to the top slide and measuring the resistance of the material between the two slides at a low shear stress. By approximating the crystals as parallelepipeds of width W and length L, at a distance r_c from the center of the rheometer, the contact area can be estimated using the following equation (Shahidzadeh et al. In press):

$$A_{rect} = \frac{h}{C}\frac{G'_{app}(w)}{G'(w)}\frac{cos(\delta)}{coscos\,(\delta_{app})}\frac{12}{12r_c^2+W^2+L^2} \qquad [2.5]$$

The rheometer measures the apparent elastic modulus $G'_{app}(w)$ from the raw data, that is, the amplitude M_a of the oscillation and the phase shift δ of the oscillating torque $M(t) = M_a \sin(\omega t + \delta)$; C is a known constant which depends on the details of the configuration (for a configuration with parallel slides $C = 2h/\pi R_p{}^4$). For confined salt crystals, we expect pure elastic behavior: $\delta = \delta_{app} = 0$ and $G'(\omega) = G$, the elastic modulus of G_{NaCl} ~12 GPa, G_{KCl} ~ 6 GPa.

During the evaporation of the saline solution, a slightly negative force develops due to the capillary forces created by the presence of the solution between the two glass slides. The measured force agrees with the calculated theoretical value, according to De Gennes et al. (2004):

$$Fc = \pi R^2 \frac{2\gamma cos\theta}{h} \qquad [2.6]$$

where R is the radius of the droplet, γ is the liquid/vapor interfacial tension of the saline solution, h is the gap between the two glass slides (50 µm) and θ is the contact angle of the saline solution with the glass slide. The calculated capillary force is about 4 mN, and decreases with time, in good agreement with the experimental values obtained using the rheometer $F_c \sim -3.5$ mN; the negative value means an attraction between the two slides.

Knowing the initial volume of the drop and its initial concentration, the evolution of the supersaturation can be quantified until the precipitation of the crystal. The latter grows rapidly and fills the 50 µm gap in the first 10 s, followed by a dominant lateral growth which evolves logarithmically (Shahidzadeh and Desarnaud 2012; Naillon et al. 2017), that is, slower than the evaporation. Interestingly, in most confinement crystallization experiments, the crystals formed are hollow cubes (Becker 1905; Royne and Dysthe 2012). This morphology results from an asymmetric growth due to a concentration gradient developing parallel to the surface of the crystal in the confined film (Figure 2.5).

During the precipitation of the crystal and its growth, a positive (repulsive) force begins to develop (Figure 2.6). A large increase in force is detected when there is no more saline solution around the unconfined faces (i.e. no more lateral growth): the remaining saline solution is in the liquid film between the glass slides and the crystal. The evaporation of this film induces the formation of steps on the confined face of the crystal as well as the precipitation of microcrystallites around the main crystal, a process otherwise known as "efflorescence" or "creeping" (Shahidzadeh et al. 2008; Qazi et al. 2019).

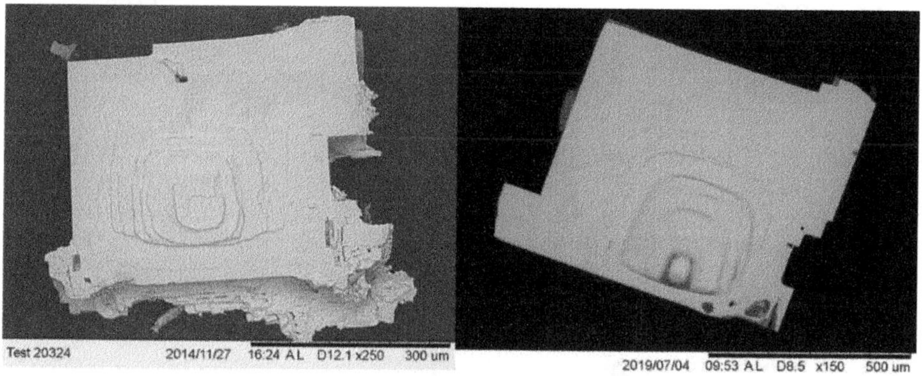

Figure 2.5. *Crystal morphologies observed by a scanning electron microscope (SEM) at the end of the evaporation experiment between two glass slides*

At the end of evaporation, the transparent appearance of the crystals under an inverted microscope makes it easy to visualize the thin films (<1 μm) of a confined solution thanks to the appearance interference colors (Figure 2.6(g)). These thin films of solution trapped between the crystalline face and the glass slide remain very stable over time (about 10 days). This type of thin film confined between the wall and the crystal face was also observed in experiments carried out in microcapillaries with sodium sulfate (Hulin 2017).

The importance of thin liquid films for crystallization pressure has already been reported theoritically (Correns 1949; Scherer 1999; Steiger 2005; Coussy 2006): for a crystallization pressure to develop, it is necessary for a liquid film to be maintained between the crystal and the wall to allow the growth of the crystal via the addition of extra ions to the crystal lattice. If a growing crystal completely fills the space between two confining walls, no crystallization pressure can develop, because no additional salt layer can be added to create such pressure.

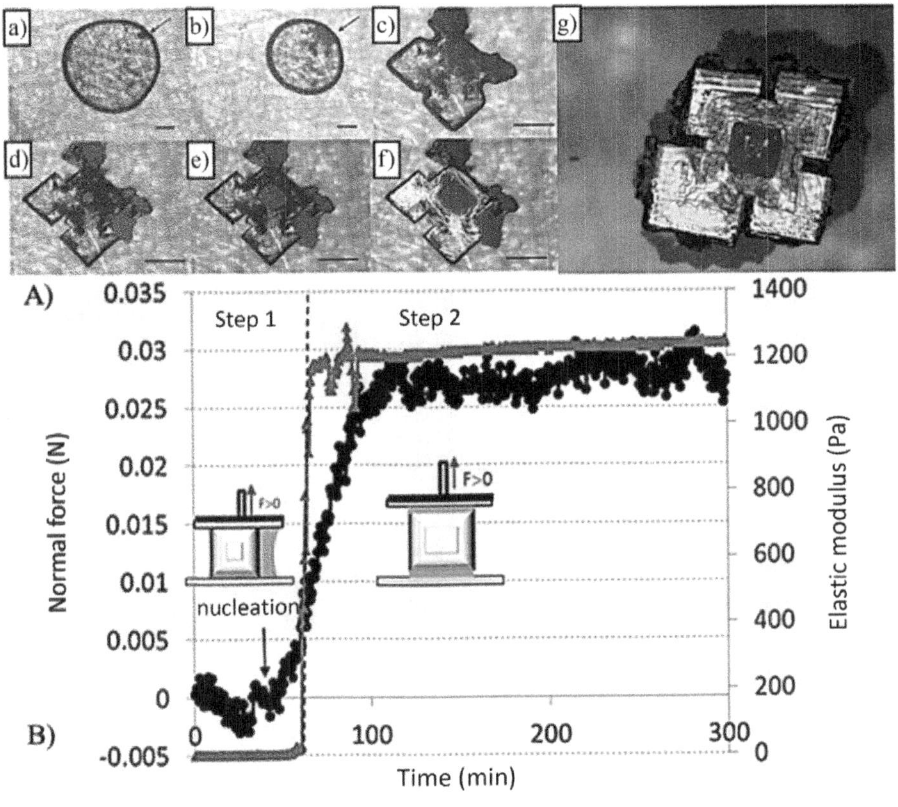

Figure 2.6. *(A) Evaporation of a drop of salt water between two slides under a microscope followed by the precipitation and growth of the NaCl crystal. (B) Simultaneous measurement of normal force and elastic modulus during crystal growth. Images (a) and (b) correspond to step 1 and (c) to (g) to step 2. For a color version of this figure, see www.iste.co.uk/derluyn/salt.zip*

The results of the different experiments are brought together by plotting the normal (repulsive) force measured as a function of the supersaturation at the time of crystallization (Figure 2.7). This concentration is the only one that we can determine with great precision. The deviation from the supersaturation value, when the crystal reaches the confinement size (in our case 50 μm) by forming a confined liquid film, can be estimated experimentally by relating the logarithmic growth rate of the crystal to the consumption of ions of the solution. Since this growth is slow (logarithmic) (Shahidzadeh 2012), the evolution of supersaturation takes place over fairly long time scales. Under our experimental conditions, the growth measurements made on several crystals after precipitation show that this difference between the concentration at the time of precipitation and when the crystal reaches the size of the confinement (the moment when a film of solution separates the two faces of the crystal of the walls) remains low, around $\Delta S \sim 0.06$ to 0.2 (Shahidzadeh et al. In press). Moreover, also due to the slow diffusion of the ions ($D_{Na+} \sim 1.3 \times 10^{-9}$ m²/s in water) during the time scale of the experiment (~1,000 s), we consider that there are few exchanges between the trapped film and the remaining solution around the side faces of the crystal.

Figure 2.7 shows that the force decreases exponentially with supersaturation; this suggests that it is the disjoining pressure Π, an internal pressure in the thin film that tends to push the interfaces apart, and is responsible for this repulsive force and the stability of the thin film (Israelashvili 1992). This pressure is then linked to a repulsion of charges between the crystal/solution and solution/glass interfaces.

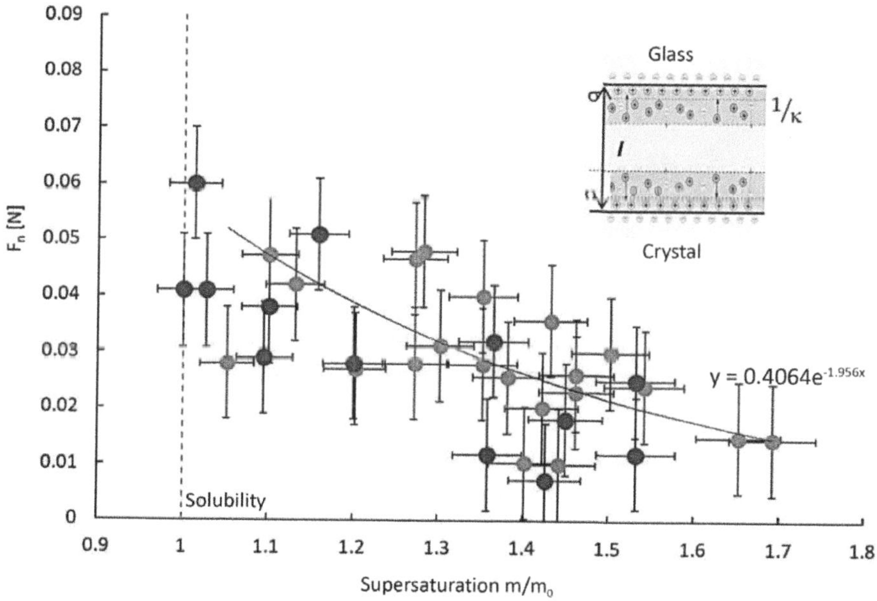

Figure 2.7. *Force (repulsive) exerted by the crystal at the end of evaporation between the two glass slides as a function of the supersaturation at the time of precipitation. For a color version of this figure, see www.iste.co.uk/derluyn/salt.zip*

The simplified form of the disjoining pressure [2.7] within a film of thickness l, pushing the two surfaces apart, can be expressed as:

$$P(l) \sim A\ exp(-\kappa l) \qquad [2.7]$$

with $A = \frac{2\sigma^2}{\varepsilon\varepsilon_0}$, where σ is the surface charge density, ε and ε_0, respectively, are the relative permittivity of the volume and the free space, $\kappa^{-1} = \frac{0.304}{\sqrt{[NaCl]}}(nm)$ is the Debye length and [NaCl] is the molar concentration of sodium chloride (Israelachvili 1992) in solution. Although it is known that glass surfaces are negatively charged when in contact with water at a low salt concentration (Siratenu et al. 2014), it is less clear whether this is still the case when they are in contact with a highly saturated salt solution (Miller and Yalamanchili 1992; Du and Miller 2007; Dihson et al. 2009).

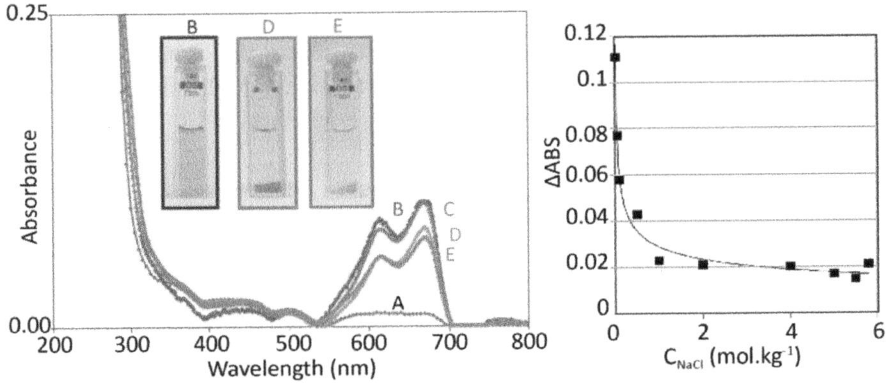

a) b)

Figure 2.8. *Characterization of surface charges. For a color version of this figure, see www.iste.co.uk/derluyn/salt.zip*

COMMENTS ON FIGURE 2.8.– *(a) Determination of surface charge by visible UV spectrometry. (A) Empty cuvette; (B) reference solution: methylene blue (MB) in 6.1 mol·kg^{-1} of NaCl reference solution in contact with (C) silanized (hydrophobic); (D) untreated (hydrophilic) glass beads; and (E) NaCl crystals added to the solution. (b) The difference in absorbance of methylene blue ABS for different solute concentrations C_{NaCl} before and after the addition of glass beads.*

The surface charges of glass and salt crystals in contact with a saturated salt solution can be measured by UV-visible spectroscopy using a cationic (methylene blue/$C_{16}H_{18}ClN_3S$, at 8×10^{-6} M) or anionic dye (eosin Y/$C_{20}H_6Br_4Na_2O_5$, at 1.2×10^{-5} M)

(Desarnaud et al. 2016). The nature of the charge of a surface and its charge density can thus be determined by measuring the discoloration of the solution due to the adsorption of the dye on the surfaces of opposite charges (i.e. the reduction absorbance due to the decrease of free molecules in the solution). The experimental results show that the glass beads and the crystal grains of NaCl and KCl are all negatively charged since they adsorb the cationic dye (Figure 2.8). On the other hand, no change is observed after the addition of hydrophobic glass beads. This measurement explains why no repulsive force is detected during crystal growth between hydrophobic glass slides (Figure 2.9).

More quantitatively, these measurements also make it possible to estimate the charge density at the surface of the silica in contact with water as well as for the solutions of NaCl and KCl. The surface charge of glass beads in water is $\sigma \sim 0.22$ C·m^{-2}, consistent with the values reported in the literature (Israelachvili 1992). The charge density of glass decreases on contact with saline solutions: $\sigma \sim 0.12$ C·m^{-2} for NaCl and $\sigma \sim 0.072$ C·m^{-2} for KCl. A lower charge density for KCl solution compared to NaCl is probably related to the larger size of the K^+ cations (ionic radius of 1.33 Å) relative to Na^+ (0.96 Å). This results in lower hydrated ionic radii. Due to a smaller hydrated radius, the K^+ ions bind more readily to the negatively charged silica surface at high concentrations (Miller and Yalamanchili 1992). As the surface charge density appears in the prefactor of the disjoining pressure expression, a much weaker repulsive force is expected for the KCl solution than for the NaCl one. Indeed, if we repeat the force measurement experiments with the KCl solution, we observe a much weaker repulsive force ($F \sim 8$ mN) (Figure 2.9).

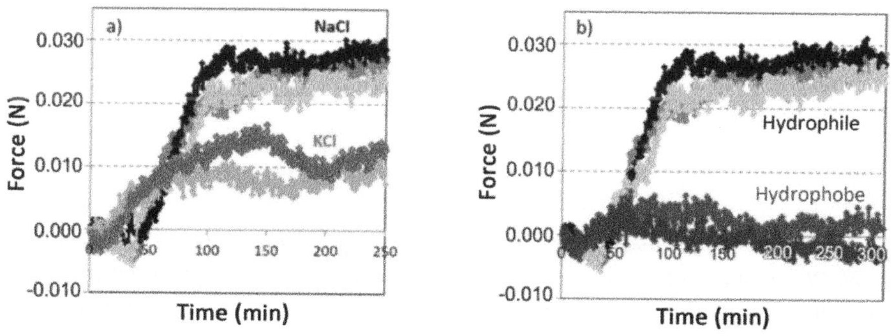

Figure 2.9. *Comparison of the detected normal force and reproducibility of measurements for (a) growth of KCl and NaCl crystals between two hydrophilic glass slides; (b) NaCl between hydrophilic and hydrophobic slides. For a color version of this figure, see www.iste.co.uk/derluyn/salt.zip*

2.4. Relationship between crystallization pressure and disjoining pressure

It is interesting to compare the thermodynamic crystallization pressure with the disjoining pressure (Scherer 2000; Desarnaud et al. 2016; Flatt et al. 2017) during crystal growth. The disjoining pressure opposes the contact between the crystal and the wall by maintaining the liquid film. The disjoining pressure therefore sets an upper limit to the crystallization pressure (Scherer 2000; Flatt et al. 2017). The latter is therefore limited by the disjoining pressure: if the supersaturation at the time of growth is high, so that $P_C > P_d$, then the crystal will be in contact with the wall and the growth of the crystal will stop.

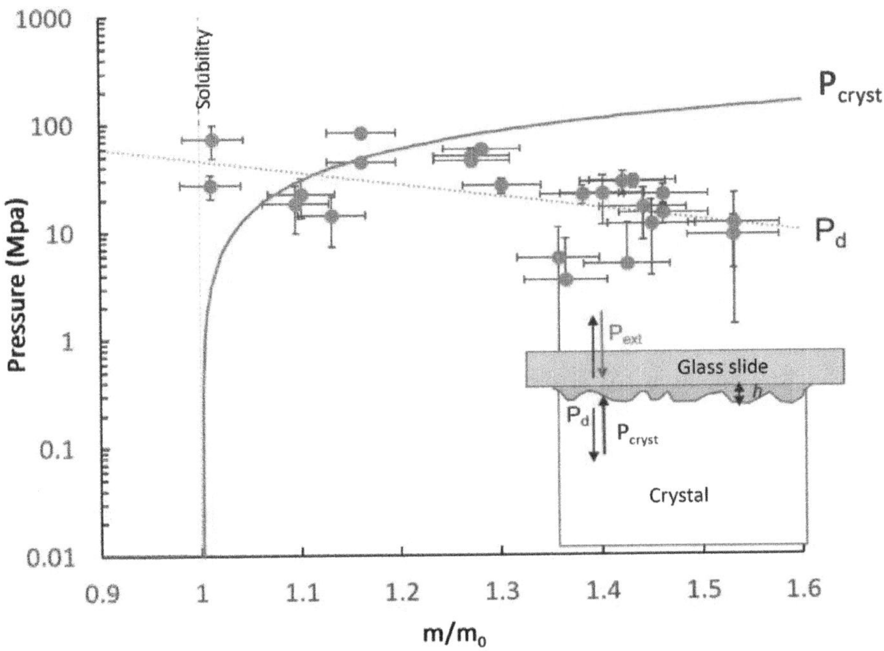

Figure 2.10. *Pressure measured (at the end of evaporation) with the precipitation of the NaCl crystal between two hydrophilic glass slides as a function of the supersaturation at the time of precipitation. The dotted line corresponds to an exponential adjustment according to the expression of the disjoining pressure P_d between two charged surfaces. The solid red curve represents the crystallization pressure calculated according to equation [2.4]. For a color version of this figure, see www.iste.co.uk/derluyn/salt.zip*

Figure 2.10 presents the experimental and theoretical results for NaCl. For each experiment, the pressure was estimated knowing the normal force at the end of the evaporation process (Figure 2.6) and the contact area of the crystal according to expression [2.5]. The equilibrium crystallization pressure calculated according to expression [2.4] is also shown in this graph. The amplitude of the two pressures remains of the same order of magnitude with, respectively, a decreasing and increasing trend. For supersaturations $S < 1.1–1.3$, due to a higher disjoining pressure than the crystallization pressure, the liquid film is maintained. Therefore, the crystal can grow and exert crystallization pressure. On the other hand, for high supersaturations ($S > 1.1–1.3$), the calculated crystallization pressure [2.4] will be greater than the measured disjoining pressure at a given supersaturation.

Figure 2.10 makes it easier to see the trend for both pressures, but this should be considered with caution since this reasoning only depends on thermodynamic equilibrium. Therefore, It does not explain why liquid films persist at high supersaturations, whereas the graph shows that the crystallisation pressure must be much higher than the disjoining pressure, and therefore the liquid films should be unstable. Nevertheless, they are observed experimentally under non-equilibrium conditions. Two hypotheses can then be formulated to explain this discrepency: (1) to determine the disjoining pressure, we make independent measurements of force and contact surface, which are both very delicate; a slight underestimation of the disjoining pressure P_d would then result in the situation $P_d = P_c$ for higher supersaturations; (2) an overestimation of the difference with the supersaturation value at the moment the crystal reaches the confinement size (ΔS). If during crystal growth, this difference drops significantly, as demonstrated for pore sizes 10 times smaller (Naillon 2018), the situation where $P_c > P_d$ could therefore move to the state where, to the state where $P_d > P_c$ and therefore the film will persist as the crystal grows over time.

2.5. Concentration in the thin films trapped between the crystal and the wall at the end of evaporation

At the end of our evaporation experiments between hydrophilic surfaces, thin films of solution remain trapped locally between the crystalline face and confined wall (glass slide). Iridescences (thin film interference patterns) are observable in these areas (Figure 2.11) and indicate a film thickness < 1 μm. Note that these interference colors do not develop when the crystal grows between two hydrophobic surfaces.

As these films remain stable for several days, we then assume that the confined film is in a state of equilibrium with the external RH (~45%) as well as with the crystal; the crystal being able to dissolve or precipitate to balance the concentration of the solution.

Figure 2.11. *(a) and (b) Iridescence of thin liquid films the end of evaporation. (c) Formation of many dendritic mini crystals (once the crystal removed); such crystal growth kinetics have been reported for supersaturations greater than 1.8 (Desarnaud 2016). For a color version of this figure, see www.iste.co.uk/derluyn/salt.zip*

Using the existing linear relationship between the equilibrium RH on top of saline solutions and the NaCl concentration in the solution ($RH_{eq} = -4.0561 \cdot [m_{NaCl}(\mathrm{mol \cdot kg^{-1}})] + 100.92$) (Pepela and Dunlop 1972), the concentration in these films is then estimated at ~13.80 $\mathrm{mol \cdot kg^{-1}}$, or a supersaturation of $S \sim 2.2$. This estimate, which seems very high, represents an upper limit, because in this estimate we do not consider the Kelvin effect (impact of the curvature of the interface on the equilibrium RH). However, the effect of curvature only becomes important for films with thickness less than 50 nm (Brinker 1990). Using the iridescences, we estimated the thickness of these films (Figure 2.9) between 100 and 700 nm, depending on the experiments. Note that the film is a liquid confined by capillarity between hydrophilic surfaces. The pressure within the solution is therefore lower than the pressure outside. On the other hand, if the crystal is under stress, its chemical potential increases. To bring the crystal/solution assembly back to equilibrium, the strained side of the crystal will therefore dissolve and increase the concentration of the film. It has also been demonstrated recently in crystallization experiments in microcapillaries (Hulin 2019) how the confinement of a salt solution by capillarity on a small scale could lead to supersolubility. Subsequently, the very slow evolution over time (several months) of this metastable state toward a state of equilibrium can generate a final stage of salt precipitation locally.

2.6. Conclusion

In this chapter, we have discussed the concept of crystallization pressure and its relation to disjoining pressure. We have detailed direct measurements of the force exerted by the spontaneous precipitation of a salt microcrystal during evaporation in confinement, like that found in porous materials. The results obtained for the growth of KCl and NaCl crystals between hydrophilic and hydrophobic glass plates suggest that the pressure arises from a repulsive interaction between charged surfaces separated by a film of saline solution. The amplitude of the measured pressure is in the same order of magnitude as the calculated crystallization pressure. The disjoining pressure therefore sets an upper limit to the crystallization pressure.

The subtle dependence of the stress generated by the wetting properties and surface charge properties may also explain the different results reported in the literature for different salts within different materials (Correns 1949; Hamilton 2010; Sekine 2011; Desarnaud et al. 2013, 2016; Naillon et al. 2018). Hydrophobic treatments such as water repellents or consolidation products have been used in the past (Scherrer 1999; Wheeler 2005); the results discussed in this chapter show how a hydrophobic treatment, removing surface charge and wetting film from the aqueous solution, could in principle reduce the development of the crystallization pressure when salt crystallizes in the pore space (i.e. in confinement). While such treatments may help limit the adverse effects of salt crystallization in practice, they could also have unintended consequences with respect to other weathering mechanisms. It should be mentioned that these unintended effects are probably not due to the hydrophobic treatment itself, but rather to its heterogeneity during application. The effectiveness of hydrophobic treatments strongly depends on the depth of impregnation of the chemicals in the stone, which in turn affects the wetting properties of porous materials and alters the flow of liquid in the porous matrix (Wheeler 2005; Shahidzadeh 2014). A heterogeneous hydrophobic treatment can lead to the formation of pockets of liquid trapped in the untreated hydrophilic regions of the porous materials, potentially allowing salt to crystallize, and therefore leading to weathering. Understanding the processes that cause the alterations of porous works of art is of great importance to manage and preserve them for future generations.

2.7. Acknowledgments

This chapter brings together the work done by several students and colleagues. I would particularly like to thank Julie Desarnaud, Étienne Jambon-Puillet, Antoine Raison, Bijoy Bera, Antoine Deblais and Daniel Bonn for their contributions and scientific discussions.

2.8. References

Becker, G.F. and Day, A.L. (1905). The linear force of growing crystals. *Proceedings of the Washington Academy of Sciences*, 7, 283–288.

Brinker, C.J. and Scherer, G.W. (1990). *Sol Gel Science: The Physics and Chemistry of Sol Gel Processing*. Academic Press Inc, Cambridge.

Colette, A., Cassar, M., World Heritage Committee (2007). Climate change and world heritage: Report on predicting and managing the impacts of climate change on world heritage, and strategy to assist states parties to implement appropriate management responses. World Heritage report n. 22, UNESCO.

Correns, C.W. (1949). Growth and dissolution of crystals under linear pressure. *Discussions of the Faraday Society*, 5, 267–271.

De Gennes, P.G., Brochard-Wyart, F., Quéré D. (2004). *Capillarity and Wetting Phenomena: Drops, Bubbles, Pearls, Waves*. Springer, New York.

Derluyn, H., Moonen, P., Carmeliet, J. (2014). Deformation and damage due to drying-induced salt crystallization in porous limestone. *Journal of the Mechanics and Physics of Solids*, 63, 242–255.

Desarnaud, J., Grauby, O., Bromblet, P., Vallet, J.M., Baronnet, A. (2013). Growth and dissolution of crystal under load: new experimental results on KCl. *Crystal Growth & Design*, 13(3), 1067–1074.

Desarnaud, J., Derluyn, H., Molari, L., de Miranda, S., Cnudde, V., Shahidzadeh, N. (2015). Drying of salt contaminated porous media: Effect of primary and secondary nucleation. *Journal of Applied Physics*, 118, 114901.

Desarnaud, J., Bonn, D., Shahidzadeh, N. (2016). The pressure induced by salt crystallization in confinement. *Scientific Reports*, 6, 30856.

Dishon, M., Zohar, O., Sivan, U. (2009). From repulsion to attraction and back to repulsion: The effect of NaCl, KCl, and CsCl on the force between silica surfaces in aqueous solution. *Langmuir*, 25(5), 2831–2836.

Du, H. and Miller, J.D. (2007). Interfacial water structure and surface charge of selected alkali chloride salt crystals in saturated solutions: A molecular dynamics modeling study. *The Journal of Physical Chemistry C*, 111(27), 10013–10022.

Dziadkowiec, J., Zareeipolgardani, B., Dysthe, D.K., Røyne, A. (2019). Nucleation in confinement generates long-range repulsion between rough calcite surfaces. *Scientific Reports*, 9(1), 1–15.

Espinosa-Marzal, R. and Scherer, G.W. (2013). Advances in understanding damage by salt crystallization. *Environmental Earth Sciences*, 69, 2657–2669.

Flatt, R.J., Steiger, M., Scherer, G.W. (2007). A commented translation of the paper by CW Correns and W. Steinborn on crystallization pressure. *Environmental Geology*, 52(2), 187–203.

Flatt, R.J., Caruso, F., Sanchez, A.M.A., Scherer, G.W. (2014). Chemo-mechanics of salt damage in stone. *Nature Communications*, 5(1), 1–5.

Flatt, R.J., Aly Mohamed, N., Caruso, F., Derluyn, H., Desarnaud, J., Lubelli, B., Espinosa-Marzal, R.M., Pel, L., Rodriguez-Navarro, C., Scherer, G.W., Shahidzadeh, N. (2017). Predicting salt damage in practice: A theoretical insight into laboratory tests. *RILEM Technical Letters*, 2, 108–118.

Goudie, A. and Viles, H.A. (1997). *Salt Weathering Hazards*. Wiley, New York.

Hamilton, A., Koutsos, V., Hall, C. (2010). Direct measurement of salt–mineral repulsion using atomic force microscopy. *Chemical Communications*, 46(29), 5235–5237.

Hulin C. (2017). Couplage géochimie-géomécanique dans les milieux poreux non saturés, Tension capillaire – Pression de cristallisation. PhD Thesis, Université d'Orléans, Orléans.

Hulin, C. and Mercury, L. (2019) Capillarity-driven supersolubility in dual-porosity systems. *Geochimica et Cosmochimica Acta*, 252, 144–158.

Israelchvili, J. (1992). *Intermolecular and Surface Forces*, 2nd edition. Academic Press, London.

Lavalle, J. (1853). Recherche sur la formation lente des cristaux. *Comptes rendus de l'Académie des Sciences*, 36, 493–495.

Lubelli, B., Van Hees, R.P.J., Huinink, H.P., Groot, C.J.W.P. (2006). Irreversible dilation of NaCl contaminated lime–cement mortar due to crystallization cycles. *Cement and Concrete Research*, 36(4), 678–687.

Miller, J.D., Yalamanchili, M.R., Kellar, J.J. (1992). Surface charge of alkali halide particles as determined by laser Doppler electrophoresis. *Langmuir*, 8(5), 1464–1469.

Mullin, J.W. (2001). *Crystallization*. Elsevier, Oxford.

Naillon, A., Joseph, P., Prat, M. (2017). Sodium chloride precipitation reaction coefficient from crystallization experiment in a microfluidic device. *Journal of Crystal Growth*, 463, 201–210.

Naillon, A., Joseph, P., Prat, M. (2018). Ion transport and precipitation kinetics as key aspects of stress generation on pore walls induced by salt crystallization. *Physical Review Letters*, 120(3), 034502.

Noiriel, C., Renard, F., Doan, M.L., Grattier, J.P. (2010). Intense fracturing and fracture sealing induced by mineral growth in porous rocks. *Chemical Geology*, 269, 197–209.

Pepela, C.N. and Dunlop, P.J. (1972). A re-examination of the vapour pressures of aqueous sodium chloride solutions at 25 °C. *The Journal of Chemical Thermodynamics*, 4(2), 255–258.

Qazi, M.J., Salim, H., Doorman, C.A.W., Jambon-Puillet, E., Shahidzadeh, N. (2019). Salt creeping as a self-amplifying crystallization process. *Science Advances*, 5(12), eaax1853.

Rijniers, L.A, Huinink, H.P., Pel, L., Kopinga, K. (2005). Experimental evidence of crystallization. Pressure inside porous media. *Physical Review Letters*, 94, 75503.

Røyne, A. and Dysthe, D.K. (2012). Rim formation on crystal faces growing in confinement. *Journal of Crystal Growth*, 346(1), 89–100.

Sabbioni, C., Cassar, M., Brimblecombe, P., Lefevre, R.A. (2008). Vulnerability of cultural heritage to climate change. Report, EUR-OPA Major Hazards Agreement, Council of Europe.

Scherer, G.W. (1999). Crystallization in pores. *Cement and Concrete Research*, 29(8), 1347–1358.

Sekine, K., Okamoto, A., Hayashi, K. (2011). In situ observation of the crystallization pressure induced by halite crystal growth in a microfluidic channel. *American Mineralogist*, 96(7), 1012–1019.

Shahidzadeh, N. (2014). Effect of hydrophobization on wetting, drying and salt crystallization in porous materials. *Restoration of Buildings and Monuments*, 20(6), 385–394.

Shahidzadeh, N. and Desarnaud, J. (2012). Damage in porous media: Role of the kinetics of salt (re)crystallization. *The European Physical Journal Applied Physics*, 60(2), 24205.

Shahidzadeh-Bonn, N., Desarnaud, J., Bertrand, F., Chateau, X., Bonn, D. (2010). Damage in porous media due to salt crystallization. *Physical Review E*, 81(6), 066110.

Shahidzadeh, N., Raison, A., Jambon-Puillet, E., Desarnaud, J., Bonn, D. (In press). The pressure exerted by growing crystals in confinement.

Siretanu, I., Ebeling, D., Andersson, M.P., Stipp, S.S., Philipse, A., Stuart, M.C., Van Den Ende, D., Mugele, F. (2014). Direct observation of ionic structure at solid-liquid interfaces: A deep look into the Stern Layer. *Scientific Reports*, 4(1), 4956.

Steiger, M. (2005). Crystal growth in porous materials I: The crystallization pressure of large crystals. *Journal of Crystal Growth*, 282(3–4), 455–469.

Taber, S. (1916). The growth of crystal under external pressure. *American Journal of Science*, 246, 532–556.

Veran-Tissoires, M. and Prat, M. (2014). Evaporation of sodium chloride solution from a saturated porous medium with efflorescence formation. *Journal of Fluid Mechanics*, 749, 701–749.

Veran-Tissoires, S., Marcoux, M., Prat, M. (2012). Discrete salt crystallization at the surface of a porous medium. *Physical Review Letters*, 108(5), 054502.

Weyl, P.K. (1959). Pressure solution and the force of crystallization: A phenomenological theory. *Journal of Geophysical Research*, 64(11), 2001–2025.

Wüst, R.A. and Schlüchter, C. (2000). The origin of soluble salts in rocks of the Thebes mountains, Egypt: The damage potential to ancient Egyptian wall art. *Journal of Archaeological Science*, 27(12), 1161–1172.

3

Evaporation, Transport, and Crystallization

Marc PRAT
IMFT, CNRS, Université de Toulouse, France

3.1. Introduction

This chapter deals with the crystallization of a salt within or at the surface of a porous medium. In the case of an aqueous solution outside the porous medium, crystallization can result from a variation in temperature or from evaporation, or even from the combined action of these two phenomena.

In this chapter, we limit ourselves to the case where temperature variations are negligible and where crystallization is essentially a consequence of the evaporation occurring within or at the surface of the porous medium. In general, the objective is to predict, and if possible, model where the crystallization takes place – in which pores, at the surface or deep in the medium, at isolated points or more extensively, etc.? – as well as the impact of the latter on evaporation – does it slow down evaporation compared to pure water, or even inhibit evaporation? A related question concerns the impact of crystallization on the porous structure itself, cracking, flaking, etc., and the possible couplings between structural alteration phenomena and transport and crystallization phenomena. Although the degradation phenomena of materials due to crystallization are a very important motivation for research on the evaporation of saline solutions in porous media, these will not be discussed in this chapter. For this, one can refer to Chapters 2 and 4 of this book. The focus of this chapter is on the evaporation-transport-crystallization trio in the absence of major mechanical effects. Another simplification concerns the aqueous solution. In

Salt Crystallization in Porous Media,
coordinated by Hannelore DERLUYN and Marc PRAT. © ISTE Ltd 2024.

many practical situations, its composition is rarely simple and the solution may contain several salts, that is, several ionic species. A logic of increasing complexity, however, leads to focus first on a simple solution containing a single salt, assuming that many generic aspects can thus be studied more simply. In fact, in this chapter, we are essentially discussing the case of a solution containing only water and sodium chloride, as this salt is very often encountered in applications. This gives the case of NaCl importance beyond the generic aspects common to all saline solutions. We do, however, mention a few references where another salt or solutions containing several salts have been considered. With this framework being established, the main points addressed follow the order evaporation, transport and crystallization. We therefore begin by recalling a few elements on evaporation in a porous medium. Then we move on to the transport aspect, that is, the factors affecting the spatial and temporal evolution of the ion concentration within the porous medium or at its surface. The coupling between evaporation-transport and precipitation is finally discussed. Another important question concerns the scale of analysis or discussion. As for other situations involving a porous medium, the phenomena of interest can be studied at the scale of a single pore (Naillon et al. 2017, 2018, 2019), pore network (Sghaier et al. 2014; Veran-Tissoires and Prat 2014; Dashtian et al. 2018) or in the classic framework of the continuum approach to porous media (Bear and Bachmat 1990). It is the latter approach that is mainly considered in this chapter. However, as this approach sometimes does not lend itself well to the interpretation or analysis of some of the phenomena mentioned, we occasionally abandon it when deemed necessary.

3.2. Evaporation in porous media

3.2.1. *Two typical situations*

Despite the diversity of evaporation situations in porous media from soil evaporation (Or et al. 2013) to the drying stages involved in the development of many materials (Mujundar 2006), reviewing the literature makes it possible to identify two generic situations very often considered where transfers can be expected to be almost one dimensional, that is, along the vertical direction. We will see, however, that it can be more subtle. These situations are schematized in Figure 3.1. For both situations, evaporation occurs at the upper surface of the porous medium. The first, illustrated in Figure 3.1(a), is called the "evaporation-wicking situation". In this situation, the porous medium is in contact at its base with an aqueous solution with a fixed ion concentration. Due to the capillary effects, resulting from the curvature of the menisci at the upper surface of the porous medium, the solution is continuously pumped through the porous medium to its upper surface where water evaporates. This corresponds to the case where the capillary effects are strong enough for the porous medium to be permanently completely saturated with the solution. When this is not the case (see, for example, Noiriel et al. 2010), the evaporation front is established within the porous medium and no longer at its upper surface. We will only consider the evaporation-wicking situation where the medium remains completely saturated. The second generic situation is drying (Figure 3.1(b)). During drying, evaporation is not compensated by the ingress of liquid water, since the walls limiting the sample (excluding the upper surface) are sealed. The medium therefore gradually desaturates as a result of evaporation.

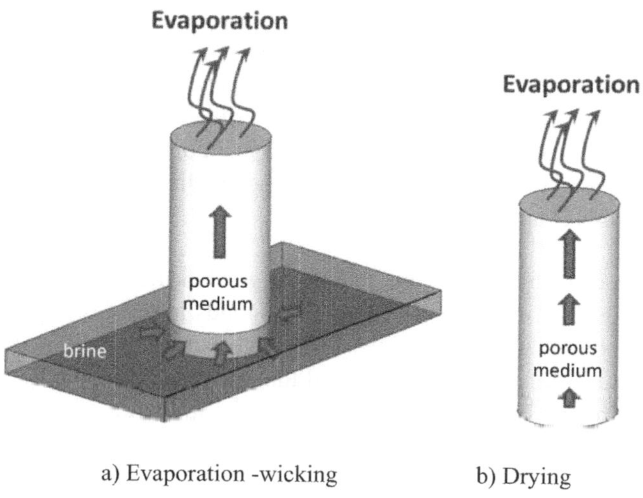

Figure 3.1. *The two evaporation situations considered in this chapter. For a color version of this figure, see www.iste.co.uk/derluyn/salt.zip*

Evaporation in the case of drying is usually characterized by the evolution of the evaporation flux j at the surface as a function of the average saturation S of the sample (S is the ratio of the total volume of solution in the porous medium to the total volume of the pore space, $S = 1$ when the medium is completely saturated, 0 when it is dry). The curve $j(S)$ is called the drying kinetics (Van Brakel 1980). Pore size has a direct impact on the drying kinetics (Van Brakel 1980).

It is usual in this context to distinguish hygroscopic media capable of absorbing a quantity of water by physical adsorption representing a significant fraction of the pore volume from capillary-porous media for which this fraction is very low. The latter typically correspond to media whose pores are larger than about 0.1–1 μm, while a highly hygroscopic medium has mainly smaller pores. We only consider the case of capillary-porous media in this chapter. The latter are generally characterized by a drying kinetics in three periods (Figure 3.2(a)). The first period is called the constant rate period (CRP) because the evaporation rate remains close in this period to the evaporation rate at the start of drying. The second period is called the falling rate period (FRP) because it is a period when the rate of evaporation decreases markedly. Finally, the last period, known as the receding front period (RFP), is characterized by a slower decrease in the evaporation rate. It corresponds to a period where the evaporation occurs entirely inside the porous medium and no longer at its surface, the FRP appearing as a transition period between the CRP, where the evaporation takes place essentially at the surface and the RFP.

In fact, the situation is more subtle and depends on the competition between capillary, viscous and gravity effects (Prat 2002). The drying kinetics in three main periods described above correspond to the situation frequently observed in laboratory experiments where capillary effects are dominant, at least over a good part of the drying. We therefore call this case the capillary regime. It is the only one we consider in this chapter.

Finally, we can note that some authors (e.g. Shokri and Or 2002) distinguish only two periods and not three. The first corresponds to the CRP and the second to the FRP-RFP combination. It can also be noted that in fact the evaporation flux often decreases a little during the CRP and that therefore designating this period as the first drying period is more appropriate than a period with a constant evaporation rate.

Despite what can be read in numerous publications devoted to the drying of capillary-porous media, there is no totally satisfactory model, that is, without fitting parameters, making it possible to predict the kinetics of drying. We can refer, for example, to Vu and Tsotsas (2018) for a presentation of the most classic model. On the other hand, a well-documented aspect in agreement with the "theory" of drying is that, in the capillary regime, the saturation profiles are (almost) spatially uniform during a good part of the CRP. This is illustrated in Figure 3.2(a). This property is utilized in the following section.

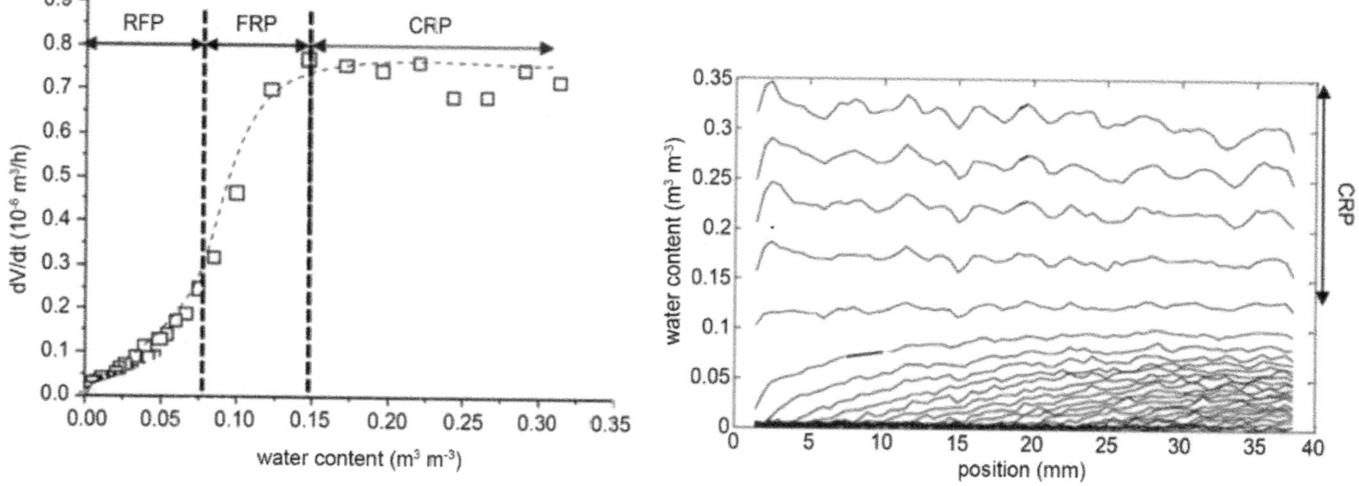

a) b)

Figure 3.2. *(a) Typical drying kinetics of a capillary-porous medium at room temperature, where V denotes the volume of water evaporated. The water content is equal to εS, where S is the water saturation and ε the porosity of the medium. (b) Saturation profiles for the capillary regime during the drying of a capillary-porous medium (according to Gupta et al. (2014))*

3.2.2. *Kinematics*

The velocity field induced in the solution by evaporation is a very important aspect due to the role of convective effects in the transport of ions in solution. In the case of the saturated evaporation-wicking situation illustrated in Figure 3.1(a), the solution is trivial when the medium is homogeneous, that is, when the porosity and permeability of the porous medium do not vary spatially. The filtration velocity is spatially uniform, directed upwards and its non-zero component equal at all points to:

$$U_z = \frac{j}{\rho} \qquad [3.1]$$

where j is the evaporation flux (kg/m^2/s) and ρ is the density of the solution. For the drying configuration in Figure 3.1(b), the velocity in the CRP varies linearly with the position along the vertical (Huinink and Michels 2002; Moghaddam et al. 2017):

$$U_z(z) = \frac{j}{\varepsilon S \rho} \frac{z}{H} \qquad [3.2]$$

where $z = 0$ corresponds to the bottom of the sample, $z = H$ is the position of the upper surface of the sample where evaporation occurs, ε is the porosity. The saturation S is obtained from a simple mass balance (Huinink and Michels 2002):

$$S = 1 - \frac{j}{\varepsilon \rho H} t \qquad [3.3]$$

taking advantage of the fact that we limit ourselves here to the period when S is spatially uniform.

3.3. Transport of ions

Since the solution is an ionic solution, modeling the transport of dissolved salts involves looking at the transfer phenomena occurring in this type of solution. Interesting information on this topic can be found in del Rio and Whitaker (2016), where a transport equation for each ion, that is, one for Na+ and one for Cl– in our case, is considered. However, a simpler approach is adopted in many studies (Huinink and Michels 2002; Guglielmini et al. 2008; Hidri et al. 2013; Diouf et al. 2018) with the consideration of a single equation for the dissolved salt. Within the framework of the continuous approach to porous media, this amounts to considering the transport equation:

$$\frac{\partial \varepsilon \rho S C}{\partial t} + \nabla.(\rho \mathbf{U} \nabla C) = \nabla.(\rho \varepsilon S D_s^* \nabla C) \qquad [3.4]$$

where C is the ion mass fraction. In the most common situations, the Péclet number based on the pore scale is sufficiently small for the effects of mechanical dispersion (Dagan 2012) to be negligible. D_s^* is therefore the effective diffusion coefficient of the ions. The latter is a function of the saturation (Hidri et al. 2013). In the case of a saturated wick, [3.4] applies by making $S = 1$. Taking the sample height H as the characteristic spatial scale and $\frac{j}{\rho}$ as characteristic velocity, [3.4] in the case of a saturated wick can be written in a 1D dimensionless form (the variation in density with the mass fraction in ions being neglected):

$$Pe\left(\frac{\partial C^*}{\partial t^*} + \frac{\partial C^*}{\partial z^*}\right) = \frac{\partial^2 C^*}{\partial z^{*2}} \qquad [3.5]$$

where $Pe = \frac{jH}{\varepsilon \rho D_s^*}$ is the Péclet number based on the sample height ($z^* = z/H$, $t^* = t/(H\rho\, \varepsilon/j)$, $C^* = C/C_0$).

In the case of a saturated wick and for an initial mass fraction in salt that is spatially uniform and equal to the mass fraction C_0 in the solution in contact with the sample at its base, an analytical solution giving the evolution of the concentration profile can be obtained (Pel et al. 2018). This solution is in very good agreement with the profiles measured by NMR (Pel et al. 2018). As shown in Figure 3.3, it is characterized by the formation of a marked gradient zone that develops from the surface where evaporation occurs. In this gradient zone, the concentration decreases exponentially with depth (Pel et al. 2018). At each point, the concentration is maximum at the surface where it increases linearly with time. Further inside the medium, the concentration gradient becomes very small, and the ion mass fraction is essentially uniform and equal to C_0. This is the profile obtained for the situation where the Péclet number Pe is not too small such that the convective effect, combined with the fact that the ions cannot escape from the medium through its upper surface, leads to the preferential accumulation toward the surface illustrated in Figure 3.3.

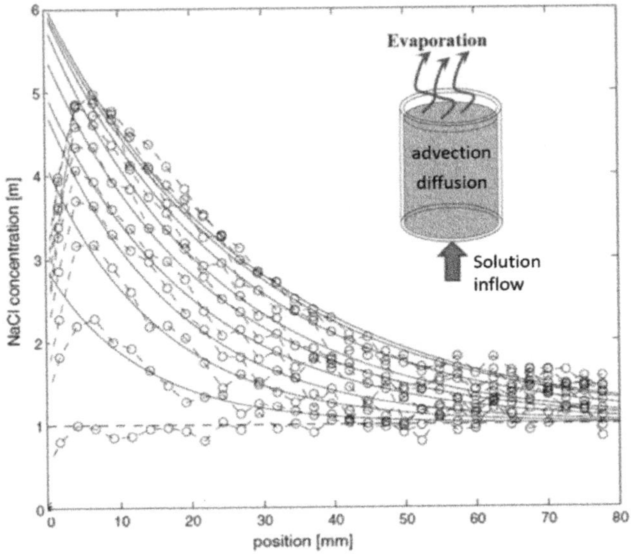

a)

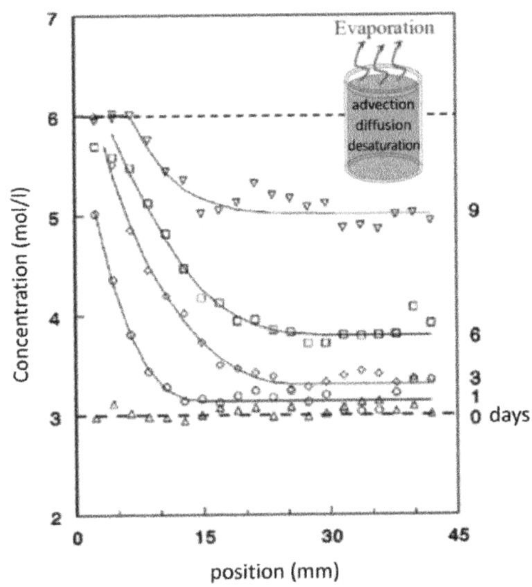

b)

Figure 3.3. *(a) Typical profiles of NaCl concentrations for a moderate Péclet number for evaporation-wicking (from Pel et al. (2018)). (b) Case of drying (according to Pel et al. (2002)). The evaporation surface is at position 0. For a color version of this figure, see www.iste.co.uk/derluyn/salt.zip*

When $Pe << 1$, convection is negligible compared to diffusion and the concentration is spatially uniform and constant everywhere and equal to C_0.

As illustrated in Figure 3.3(b), the situation is qualitatively similar in drying to the evaporation-wicking situation, with the formation of a gradient zone in the upper part of the sample when the Péclet number is not too small. However, the variation in concentration does not result only from the convective effect in the case of drying. As the saturation decreases during drying, the concentration necessarily increases on average by this effect since the quantity of dissolved salt occupies a decreasing volume of liquid during drying. During the CRP, the average salt mass fraction is given by $\bar{C}(t) = S^{-1}C_0$ (neglecting the variation in density of the solution with the ion mass fractions). In the case where $Pe << 1$, the local salt mass fraction is spatially uniform and equal to $C(t) = S^{-1}C_0$. The effect of increasing concentration due to desaturation is very clearly visible in Figure 3.3(b).

For more details for the case of drying, one can refer to some previous works (Huinink and Michels 2002; Sghaier et al. 2007; Guglielmini et al. 2008; Hidri et al. 2013; Rad et al. 2013). Unlike the case of the evaporation-wicking situation, there seems to have been no detailed comparison between experimental profiles and predictions based on [3.4]. We can however quote Shokri (2014), where in fact it is concluded that [3.4] only allows a qualitative prediction. However, the comparison presented in Shokri (2014) is for a case where only about 10 grains make up the sample height, which makes the value of the comparison doubtful given the obvious lack of scale separation (about 10 particles correspond to less than one representative elementary volume [REV]).

3.4. Supersaturation

When the salt concentration reaches a sufficient value, which we call the crystallization concentration, the salt no longer remains dissolved but instead precipitates. This is the beginning of crystallization. A question is whether this crystallization occurs at solubility (6 M for NaCl, or 0.264 in mass fraction) or for a higher value, that is, for a supersaturation > 1. This point is an important subject in relation with the concept of crystallization pressure (see Chapter 2), the latter depending directly on supersaturation. In model systems, such as capillary tubes or single channels (Desarnaud et al. 2014; Naillon et al. 2017, 2018), quite notable supersaturations (~1.7–1.8) were obtained. In contrast, estimates from profiles measured by NMR in porous samples (Pel et al. 2018) instead suggest negligible supersaturation (~1). One can argue that the NMR gives a spatially integrated measurement, and therefore cannot reliably monitor the spatial fluctuations in concentration at the surface (due to edge effect and problem of spatial resolution).

Moreover, the crystallization kinetics are very fast in the case of NaCl (Naillon et al. 2017) so that once the crystallization has started locally, the excess ions are very quickly consumed. This question is still not resolved. In other words, should we expect supersaturation levels comparable to those measured in model systems in the case of real porous media and/or in practice?

3.5. Subflorescence and efflorescence

For the reference situations presented above, the maximum concentration is at the surface of the porous medium (except for the case of very low Péclet numbers) and crystallization is therefore expected at the surface. Efflorescence refers to the clusters of crystals growing at the surface. In other circumstances, the crystallization concentration can be reached inside the medium, for example in the evaporation-wicking situation, if the conditions are such that only part of the porous medium is invaded by the solution (Noiriel et al. 2010). The subflorescence is the sets of crystals developing inside the porous medium. Figure 3.4 illustrates these two types of crystallization that formed in this example during a drying experiment with a model porous medium (Sghaier et al. 2014).

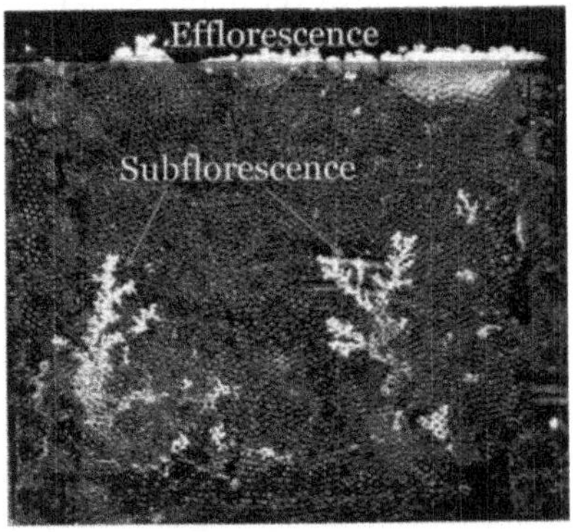

Figure 3.4. *Efflorescence and subflorescence in a model porous medium (Sghaier et al. 2014). The salt structures forming inside the porous medium are called subflorescence while efflorescence refers to the salt structures (crystals) on the surface of the porous medium. For a color version of this figure, see www.iste.co.uk/derluyn/salt.zip*

A very important point is that these structures are porous (Sghaier and Prat 2009; Veran-Tissoires and Prat 2014). This is a key element to analyze their growth and their impact on evaporation.

3.6. Branching efflorescence versus compact efflorescence (crust)

An important characteristic of the development of efflorescence (and subflorescence) is that its morphology can vary significantly depending on the conditions. A rough classification consists of distinguishing two main types (Eloukabi et al. 2013): branching and compact efflorescence (Figure 3.5). Branching efflorescence can also be called cauliflower-like or discrete efflorescence. The latter name (Veran-Tissoires et al. 2012a) comes from the fact that the efflorescence often does not completely cover the surface of the porous medium. The glass beads forming the porous medium are visible between the efflorescence in Figure 3.5(a).

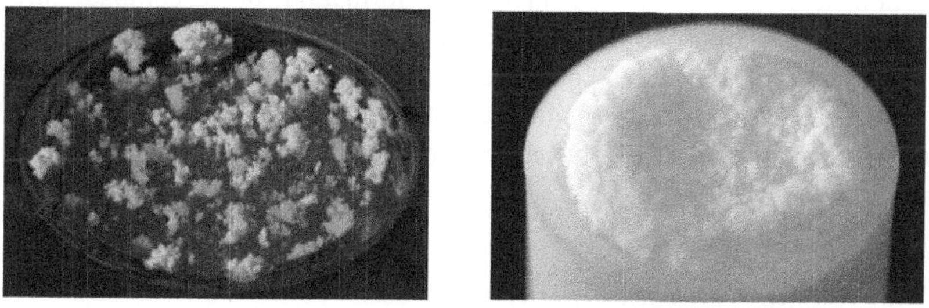

a) b)

Figure 3.5. *(a) Branching efflorescence. (b) Compact efflorescence*

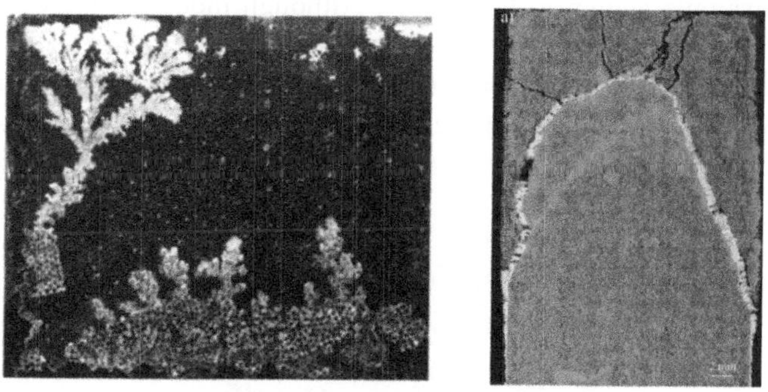

a) b)

Figure 3.6. *(a) Branching subflorescence observed in a model system with large pores (about 100 μm) (Sghaier et al. 2014). (b) Image of compact subflorescence in a porous rock obtained by X-ray microtomography (Noiriel et al. 2010)*

Compact efflorescence is also called crust. As shown in Figure 3.5(b), this type of efflorescence tends to grow by covering the surface of the porous medium. As can be seen, the crust-like efflorescence shown in Figure 3.5(b) completely covers the upper surface of the porous medium. Unlike the efflorescence in Figure 3.5(a), the underlying porous substrate is not visible in Figure 3.5(b). In the case of subflorescence, one can also observe branching or compact subflorescence depending on the conditions. This is illustrated in Figure 3.6.

Schematically, large pores (over 100 μm) and rather low evaporation rates (rather high relative humidity [RH] in the ambient air) favor the appearance of branched structures, whereas a porous substrate with smaller pores and higher evaporation rates (e.g. dry air) favor the formation of a compact structure. Understanding the morphology transition with pore size and/or evaporation conditions, however, remains a completely open topic. We can find in the literature some considerations on this transition and the beginnings of an explanation (e.g. Veran-Tissoires and Prat 2014), but much remains to be done in order to understand this transition in detail.

3.7. Efflorescence entrenchment

As seen in Figure 3.4 and illustrated in Figure 3.7, efflorescence is not a strictly superficial structure. Precipitation occurs at the surface of the porous medium but also within the porous medium over a distance of a few pore sizes, one to three grain sizes in the example shown in Figure 3.7. Although modeling the development of efflorescence is still a rather underexplored subject (Licsandru et al. 2019), this element is certainly to be considered in the models, particularly in connection with the analysis of the impact of efflorescence on the drying kinetics as discussed later in this chapter.

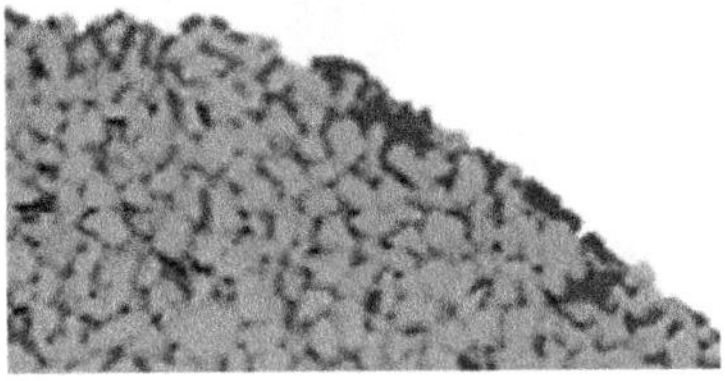

Figure 3.7. *Image of compact efflorescence (in red) on the surface of a porous rock obtained by X-ray microtomography (Desarnaud et al. 2015). For a color version of this figure, see www.iste.co.uk/derluyn/salt.zip*

3.8. Factors affecting surface efflorescence distribution

We can distinguish two stages in the formation of efflorescence, the "nucleation" stage (in reality, the crystals are generally observed in experiments only when they have reached a size clearly larger than the initial germs) and the growth stage. A question of interest is to predict where the first crystals will form at the surface of the porous medium. As illustrated in Figure 3.8, this issue is not trivial even for the simple configurations considered in this chapter.

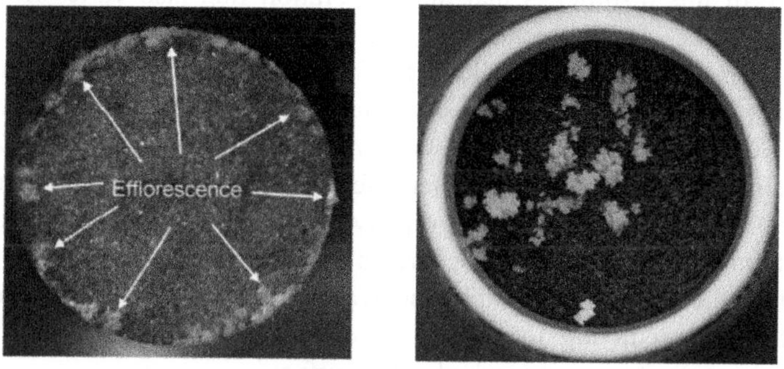

a) b)

Figure 3.8. *(a) Fairy ring: discrete efflorescence forming on the edge of the upper surface of the sample. (b) Discrete efflorescence without peripheral effect*

As seen in this example, the (discrete) efflorescence can form on the periphery (this structure is called a "fairy ring" by purely visual analogy to fairy rings, also called witches' rings, that mushrooms can form) or, on the contrary, without any noticeable peripheral effect. As analyzed in detail in Veran-Tissoires and Prat (2014) for the evaporation-wicking situation and in Hidri et al. (2013) for drying, the fairy ring structure is a consequence of the higher evaporation rate at the periphery (and therefore according to [3.1] of a convective transport effect induced in the porous medium, which is stronger at the periphery). The distribution of the evaporation rate (Veran-Tissoires et al. 2012a; Veran-Tissoires and Prat 2014) is in this case analogous to the classic coffee-ring effect in the case of the evaporation of a droplet deposited on a surface (Deegan et al. 1997). On the contrary, in the case of the experiment illustrated in Figure 3.8(b), the distribution of the evaporation flux is much more uniform over the surface and the peripheral effect is no longer present. Thus, a rule emerges. In the case of a homogeneous medium, the sites at the surface of the porous medium where evaporation is stronger are preferential places of efflorescence formation.

The discrete aspect of the efflorescence in Figure 3.8 cannot be analyzed using the continuum approach, that is, from [3.4], because it involves fluctuations in ion concentrations at a smaller scale than the R.E.V. As discussed in detail in Veran-Tissoires and Prat (2014), there are two main sources of fluctuations in concentration, both related to the disordered structure of the porous medium (the fact that the pore sizes are distributed over a certain range). The first is due to the convective transport of ions. As the velocity varies spatially from pore to pore due to disorder, this leads to concentration fluctuations in the pore network. Another consequence of this disorder is that the evaporation flux also varies from pore to pore at the surface. These fluctuations lead to spatial velocity fluctuations in the underlying pores, which in turn lead to fluctuations in ion concentration. The "nucleation" points are then interpreted as sites of the local concentration maxima at the surface just before crystallization starts. As discussed in Hidri et al. (2013) and Veran-Tissoires and Prat (2014), another effect can also have an impact in the case of granular media: the locally greater porosity near the walls limiting the sample due to the particular arrangement of the grains in the vicinity of a wall. Thus, the problem of predicting the most likely "nucleation" sites on the surface can be seen as a problem of predicting local maxima of ion concentrations just before the crystallization concentration is reached at the surface.

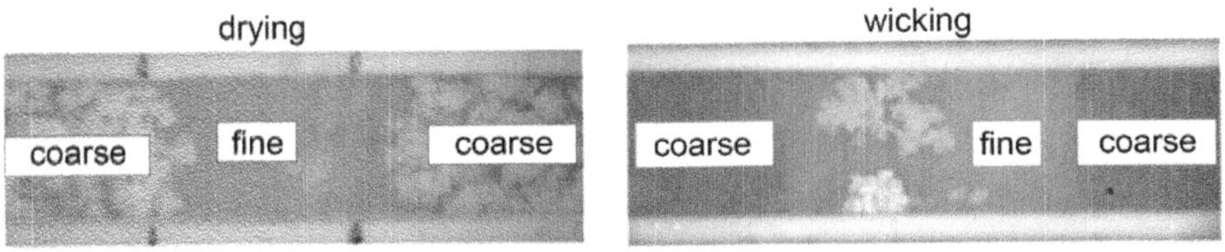

Figure 3.9. *Impact of large-scale heterogeneity on the location of efflorescence formation. Case of a heterogeneous medium formed by a column of fine medium placed between two columns of coarse medium (Diouf et al. 2018)*

In the case where larger-scale heterogeneities are present (media where the porosity and/or permeability vary spatially), it is these that will generally control the places where crystallization will begin. This is illustrated in Figure 3.9 in the case of columnar heterogeneities (Diouf et al. 2018). It can be seen in Figure 3.9 that the efflorescence develops on the surface of the region of lowest permeability (fine medium) in the case of the evaporation-wicking situation, whereas exactly the opposite is observed in the case of drying. As explained previously, the change in concentration is essentially dependent on the velocity field in the case of the evaporation-wicking situation, whereas in the case of drying, the variation in concentration also depends on the variation in saturation. A consequence of capillary effects is that during drying, the coarse medium desaturates preferentially while the fine medium remains saturated. As analyzed in Diouf et al. (2018), this preferential desaturation explains the development of efflorescence at the surface of the coarse medium during drying. In the case of the evaporation-wicking situation, it is the structure of the velocity field (preferential flow in the coarse medium with redirection toward the fine medium near the surface) that explains the formation of efflorescence at the surface of the fine medium. In the case of the evaporation-wicking situation, this result generalizes to more complex permeability variations (Hidri et al. 2019) with a correlation between local concentration extrema and stagnation points of the velocity field at the surface. Further illustrations of the effect of large-scale heterogeneities on the sites of crystallization can be found in some previous studies (Bechtold et al. 2011; Nachshon et al. 2011a, 2011b; Veran-Tissoires et al. 2012b; Bergstad et al. 2017, 2018).

3.9. Efflorescence dynamics

Efflorescence dynamics, that is, their evolution after the “nucleation“ stage, is linked to their porous structure. The growth phenomenology is illustrated in Figure 3.10(a) for a branched efflorescence and in Figure 3.10(b) for a compact efflorescence.

The phenomenology is as follows: since the efflorescence is a porous structure, the solution is pumped by capillarity through the efflorescence up to the efflorescence periphery where the water evaporates, and the salt contained in the solution precipitates to form a new crystal layer. Thus, we expect a direct link between the local growth rate of the efflorescence and the local evaporation flux. This correlation has been formally established (Sghaier et al. 2014) for the subflorescence shown in Figure 3.6(a), whose growth is analyzed in a similar way.

Thus, for the case illustrated in Figure 3.10(b), this suggests a very simple growth law depending on the evaporation flux from a simple mass balance:

$$\frac{dz_t}{dt} = \frac{C_{sat}\, j}{\rho_{cr}(1-\varepsilon)(1-C_{sat})} \qquad [3.6]$$

where z_t is the position of the upper surface of the efflorescence (for simplicity, we consider a point on the surface where the growth takes place along the vertical), C_{sat} is the equilibrium mass fraction in a saturated solution ($C_{sat} = 0.264$), ε is the crust porosity and ρ_{cr} is the density of the crystal ($\rho_{cr} = 2{,}160$ kg/m^3).

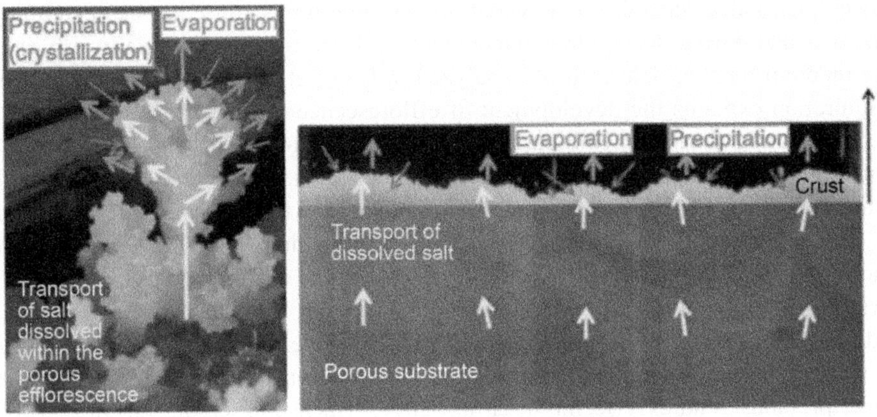

a) b)

Figure 3.10. *Illustration of the main mechanisms involved in efflorescence growth: (a) branched efflorescence; (b) crust efflorescence. For a color version of this figure, see www.iste.co.uk/derluyn/salt.zip*

In fact, as discussed in Licsandru et al. (2019), relation [3.6] neglects the very slight supersaturation effect necessary for the salt to precipitate. When this effect is considered, we obtain a growth law in the form:

$$\frac{dz_t}{dt} = \frac{C_{sat}\, j}{\rho_{cr}(1-\varepsilon)(1-C_{sat})}\left[\frac{Da}{1+Da}\right] \qquad [3.7]$$

where $Da = \sqrt{\frac{(1-\varepsilon)^2(1-C_{sat})k_r}{\varepsilon D_s^* a_v}}$ is a Damkhöler number characterizing the competition between the precipitation reaction and the transport of ions by diffusion (Mullin 2001). In this expression, k_r is the coefficient of the precipitation reaction (Mullin 2001; Naillon et al. 2017), a_v is the specific area of the efflorescence and D_s^* is the effective diffusion coefficient in the efflorescence. Equation [3.7] explicitly shows that the growth results from the coupling between the internal transport phenomena of the ions within the efflorescence – in this example only the diffusion – and the external transfer phenomena controlling the evaporation flux j at the efflorescence surface.

3.10. Impact of efflorescence on drying kinetics

The impact of the development of efflorescence on the drying kinetics or on the evolution of the evaporation flux in the case of the evaporation-wicking situation depends on the type of efflorescence. When efflorescence is branched, the evaporation flux is little affected (Elkouabi et al. 2013; Veran-Tissoires and Prat 2014; Shokri-Kuehni et al. 2017a); it can even increase as the result of the growth of efflorescence in the external air (Sghaier and Prat 2009). The efflorescence can then grow by several millimeters in height without having a very noticeable impact on evaporation. The situation is quite different in the case of crust-like efflorescence developing on a large fraction of the open surface of the porous medium or its entirety. In this case, the evaporation flux can be considerably reduced compared to its initial value when the surface was free of efflorescence. As mentioned in section 3.7, a factor affecting the type of efflorescence that forms is the level of the initial evaporation flux, that is, before the efflorescence is formed. A high flux, with all other things being equal, favors the formation of a salt crust, while a lower flux can lead to branching efflorescence. This can lead to the paradoxical evolution presented in Gupta et al. (2014). In the case of pure water, the drying is faster as the RH decreases in the ambient air; an opposite situation is reported in Gupta et al. (2014). Due to the change in efflorescence morphology (crusted or branched), drying is slower at low RH than at higher RH in the case of evaporation with a NaCl solution for the conditions considered in Gupta et al. (2014).

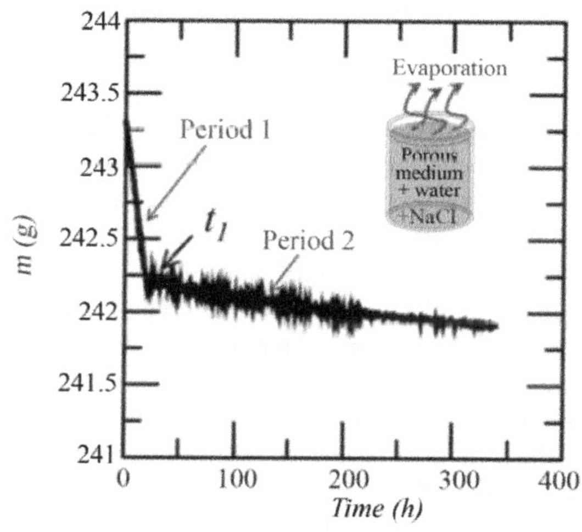

a)

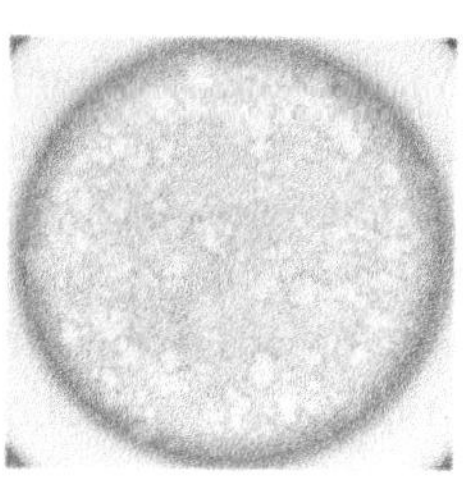

b)

Figure 3.11. *(a) Variation in the mass of a cylindrical porous medium sample (diameter of glass beads in the range [1–50 µm]) initially saturated with a solution of NaCl close to solubility during drying at room temperature (RH = 7%). (b) Crust completely covering the upper surface of the sample (period 2). For a color version of this figure, see www.iste.co.uk/derluyn/salt.zip*

Figure 3.11 shows a typical example when the development of the crust strongly impacts the evaporation kinetics (Duenas-Velasco 2016). In this example, the crust does not completely prevent evaporation but very much reduces it. The flux in period 2 is about 50 times lower than in period 1. Period 1 corresponds to a period when the crust develops without affecting the evaporation flux. In period 2, the crust completely covers the upper surface of the porous medium (Figure 3.11(b)).

A key question is obviously the explanation of this very strong reduction in evaporation flux. Like many other questions in this area of research, it has still not been resolved and is in fact an open topic. In Desarnaud et al. (2015), the crust is dry during period 2, so that the reduction in flux is then entirely explained by the impact of the crust as a vapor diffusion barrier. Since the crust is assumed to have low porosity, its vapor effective diffusion coefficient would be low enough to reduce the flux in the proportion observed, despite the low thickness of the crust. However, very often, for example the case in Figure 3.11, the very low flux period 2 is observed while the saturation in the underlying porous substrate is still very high. In other words, the crust is in contact at its base with the solution. It is established that the solution wets almost perfectly the halite (Corti and Krieger 2007). For the diffusion barrier explanation to be convincing, it would be necessary to explain why the crust remains dry under these conditions. On the contrary, it is expected that it remains at least partially saturated in solution by capillary action. An alternative approach is to assume that the crust stays wet. The solution is transported through the crust by capillarity to its upper surface where the salt precipitates. The pattern in terms of transport and precipitation phenomena is analogous to that shown in Figure 3.10(b). One element in favor of this scenario is that the upper surface of the crust moves upwards during the low evaporation period 2 (Duenas-Velasco 2016). Considering that this shift is purely transport related (see some additional discussion below), the only mechanism to explain the observed shift is salt precipitation on the upper surface of the crust. In this scenario, salt precipitation leads to progressive pore plugging and therefore changes the pore distribution at the upper surface of the crust. Based on the classic Schlünder model (Talbi and Prat 2019) describing evaporation from a perforated surface, it then seems possible to explain the flux reduction by the reduction in surface porosity on the upper surface of the crust. In fact, the analysis presented in Licsandru et al. (2022) suggests that the pores on the surface of the crust become small enough for the Kelvin effect to have an impact. Finally, another argument presented in Nachshon et al. (2018) is based on the observation that the crust can in fact be detached from the surface of a porous medium (Figure 3.12). A gap between the bottom of the crust and the surface of the porous substrate of approximately 1 mm is mentioned in Nachshon et al. (2018).

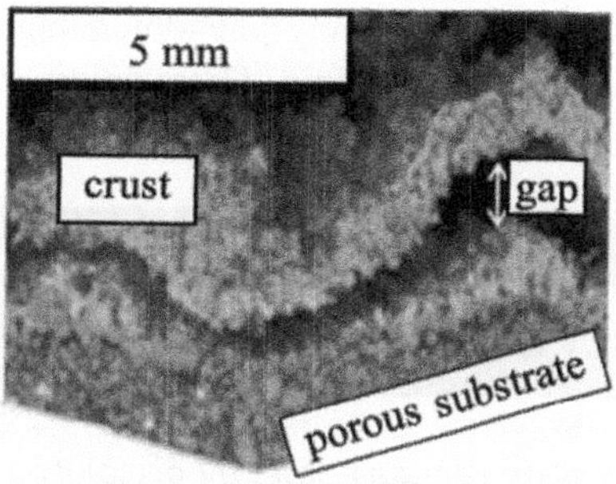

Figure 3.12. *Example of crust detached from the porous substrate on which it initially developed (according to Nachshon et al. (2018)). For a color version of this figure, see www.iste.co.uk/derluyn/salt.zip*

Under conditions where the gap is greater than the capillary length, the scenario based on the crust as a barrier to vapor phase diffusion regains validity. It is reasonable to imagine that a crust detached from the surface is dry. However, this scenario is problematic because it does not explain how the crust detaches and we do not know when this detachment of the crust occurs from the transition point between the two periods (point t_1 on Figure 3.11(a)) or much later in period 2.

Two mechanisms likely to explain this detachment are briefly mentioned in Nachshon et al. (2018). The first invokes possible mechanical effects in connection with the fact that crystallization can have mechanical consequences (see Chapters 2 and 4 of this book). On the other hand, another approach involves neglecting any effect of mechanical origin and links the detachment to dissolution-precipitation effects.

As illustrated in Figure 3.13, the fact that the displacement of a crust can be fully explained by dissolution-precipitation mechanisms was clearly established from an experiment with a salt crust set in a Hele-Shaw cell (Licsandru et al. 2019; Licsandru 2020). However, this does not provide a definitive conclusion regarding possible mechanical effects, especially in field studies. It is clear however that dissolution-precipitation phenomena are key elements for understanding the dynamics of salt crusts (see also Licsandru et al. (2022, 2023)), and more broadly efflorescence and subflorescence.

In summary, the missing pieces in order to explain and model outcomes such as the one shown in Figure 3.11 are now in place. However, significant research is still required, particularly in terms of numerical simulations (see section 3.11) or characterization (structural properties of the crusts), to clearly explain this outcome.

A practical alternative is to develop purely phenomenological models to reproduce the experimental results (see, for example, Grementieri et al. (2017)). It is clear though that a detailed physical approach is the desired objective.

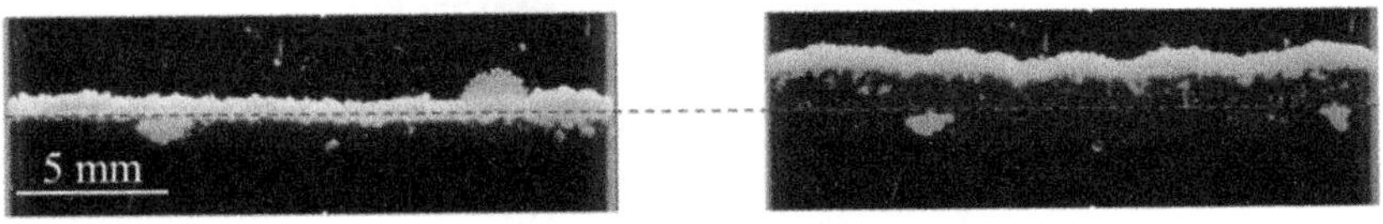

Figure 3.13. *Upward migration of a salt crust by dissolution on its lower surface and precipitation on its upper surface (Licsandru et al. 2019). The time elapsed between the left image and the right image is 18 days. For a color version of this figure, see www.iste.co.uk/derluyn/salt.zip*

3.11. *Patterned crusts*

Salt crusts can have highly organized structures. A famous example is the polygonal cell structure on the surface of the Salar de Uyuni in Bolivia (Figure 3.14(a)).

a) b)

Figure 3.14. *(a) Polygonal cells observed on the surface of the Salar de Uyuni in Bolivia. Pattern size is metric. (b) Salt crust with cell structure formed in the laboratory (Veran-Tissoires and Prat 2014) by evaporation of an aqueous solution of NaCl saturating a porous medium (the diameter of the cylinder is 38 mm). For a color version of this figure, see www.iste.co.uk/derluyn/salt.zip*

Given their resemblance to Bénard cells (Bénard 1901), it is tempting to link this structuring to possible thermosolutal convection effects in the underlying porous layers. Without questioning the possible impact of these effects in a natural situation (Lasser et al. 2020), quite similar structures, but with centimetric rather than metric cells, have been observed in the laboratory (Figure 3.14(b)). In this experiment, the impact of natural convection effects within the porous medium is totally negligible. The structuring is then completely linked to the growth of the efflorescence from "nucleation" points randomly distributed on the surface. This distribution is explained by the analysis of spatial fluctuations in ion concentration at the surface during evaporation. How it grows from these seeds, however, is not understood. This polygonal pattern structuring issue is another example of a poorly understood area of research. The same is true regarding the scale of the patterns, which are very different in the two images in Figure 3.14.

3.12. Evaporation and formation of subflorescence

Similar questions about the impact of efflorescence on evaporation exist regarding subflorescence. For example, does subflorescence growth affect evaporation? It seems that this question has been much less studied in comparison. Here again, we expect a different impact depending on whether the subflorescence is branched (Figure 3.6(a)) or compact (Figure 3.6(b)). The experiment on a model medium presented in Sghaier et al. (2014) with a branched subflorescence (Figure 3.6(a)) indeed leads to a similar result as with branched efflorescences. That is, a slight increase in the evaporation flux with the subflorescence growth toward the open surface of the model medium (upper edge in Figure 3.6(a)), which is explained by the "wet" nature of the subflorescence. The subflorescence growth brings the internal evaporation front closer to the surface. Similar results are observed with sand columns (Nachshon and Weisbrod 2015), which is not surprising given the size of the grains of sand used (about 400 μm). The situation where subflorescence growth severely limits the evaporation flux (analogous to the case of compact efflorescences) was observed during experiments on sandstone samples (pore size ~1–10 μm) with potassium chloride (Peysson 2012). Here, we assume a comparable phenomenology is observed with NaCl if the pores are small enough to lead to a compact subflorescence.

3.13. Models

With regard to modeling, limited to the continuum approach to porous media, one type of models has the objective of modeling the transport of ions in the presence of evaporation, either for drying or the evaporation-wicking situation (Huinink et al. 2002; Sghaier et al. 2007; Guglielmini et al. 2008; Hidri et al. 2013; Diouf et al. 2018). This type of model typically couples equation [3.4] with the classical equations used to model drying or flow in a wicking situation. An additional step involves taking into account the precipitation phenomena (and dissolution if necessary) (see, for example, Xu et al. 2006; Pruess and Müller 2009; Koniorczyk 2012; Jambhekar et al. 2015, 2016; Grementieri et al. 2017; Mejri et al. 2017). It should be noted that these models can be extended to situations where several salts are present in solution (Mejri et al. 2017). Finally, the most complete models also consider the poromechanical effects (Koniorczyk and Gawin 2012; Derluyn et al. 2014). These models model the precipitation in the pores of the porous substrate but are generally not designed to model the development of efflorescence, whether branched or crust-like at the surface of the porous medium, so in effect modeling drying with development of a crust illustrated in Figure 3.11 remains an open subject. It can be noted that a model for the transfer in the crust is proposed in Licsandru et al. (2019). It is possible to couple this type of model to more classical models describing growth in the porous substrate. An example of such a model in the case of the evaporation-wicking situation is presented in Lazhar et al. (2020) and Roy et al. (2022).

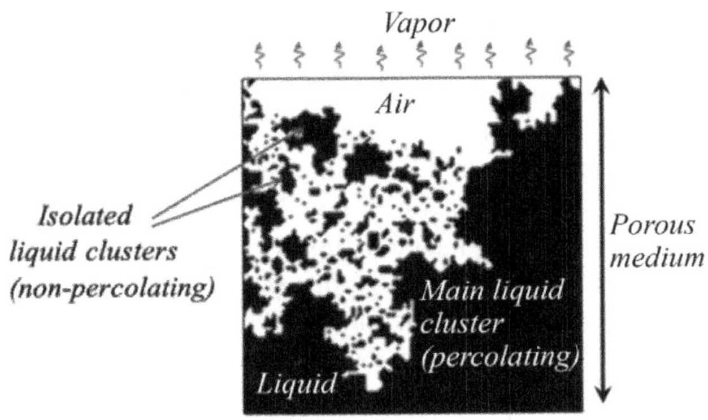

Figure 3.15. *Illustration of the liquid phase fragmentation process during drying with the main cluster (percolating liquid phase) and isolated clusters (non-percolating liquid phase). For a color version of this figure, see www.iste.co.uk/derluyn/salt.zip*

Recent work focuses on the impact of the fragmentation of the liquid phase during drying on the distribution of the solute (Ahmad et al. 2021; Talbi and Prat 2022) based on pore network simulations. As illustrated in Figure 3.15, fragmentation refers to the fact that during drying, the liquid phase fragments into many liquid clusters (Lebray and Prat 1999). This leads to a different evolution of the solute concentration in the main liquid cluster and in the isolated clusters forming the non-percolating liquid phase. The classic continuum model (Huinink et al. 2002) cannot account for this phenomenon. As shown in Talbi and Prat (2022), it is possible to take this into account by developing a two-equation continuum model considering two saturations, one for the percolating liquid phase and the other for the non-percolating liquid phase, and two equations therefore for the concentration. This allows a more precise calculation of the evolution of the concentration in the medium and the moment when the concentration reaches a given level of supersaturation (corresponding to the onset of crystallization).

Another recurring problem in drying modeling concerns the role of capillary liquid films developing along geometric singularities in the pore space. Figure 3.16 illustrates the case of liquid films developing along the four corners of a square capillary tube cross-section. It has been shown that these films contribute to maintaining a high value of evaporation thanks to their liquid phase transport capacity (Prat 2007; Chauvet et al. 2009; Yiotis et al. 2012).

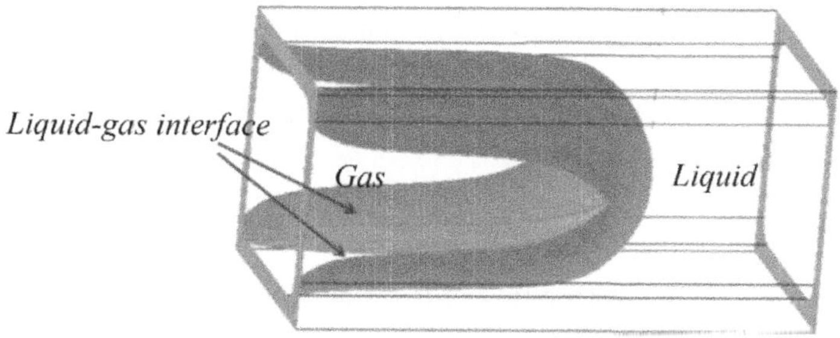

Figure 3.16. *Illustration of capillary films developing during drying (case here of a simple capillary tube of square cross-section). For a color version of this figure, see www.iste.co.uk/derluyn/salt.zip*

As discussed in Camassel et al. (2005), dissolved salt can also be transported in these films. Current models do not take this phenomenon into account, the importance and consequences of which therefore still need to be studied. It should be noted that these films can only develop if the contact angle is small enough (less than 45° in the case of the square cross-section tube illustrated in Figure 3.16). It has been observed in the case of NaCl that the presence of dissolved salt tends to increase the contact angle (Sghaier et al. 2006), which therefore does not favor the development of liquid films. Nevertheless, it is very likely that these films will have an impact in many cases. We can also think that their effect can be combined with the phenomenon of salt creeping (Qazi et al. 2019) in the pores, leading to more subtle distribution dynamics of the precipitated salt within the porous medium than considered in the current models.

3.14. Other salts

As indicated previously, we have only considered the case of aqueous solutions of sodium chloride. In addition to the fact that it is commonly encountered in practical situations, it has very high solubility compared to most other salts. This explains why the evaporation of NaCl solutions can lead to very significant efflorescence or subflorescence growth. Another differentiating characteristic is that under the usual conditions, the crystallized sodium chloride is anhydrous (a hydrated form only exists for a cold temperature, typically below 0°C). This simplifies the modeling compared to the case of hydrated salts such as sodium sulfate, which can also crystallize in different hydrated forms in addition to the non-hydrated form (Rijniers et al. 2005). Some elements of comparison between the case of sodium chloride and sodium sulfate can be found in Shahidzadeh-Bonn et al. (2010) and Shahidzadeh and Desarnaud (2012) while a comparison between the case of NaCl, calcium chloride and potassium iodide is presented in Shokri-Kuehni et al. (2017). Material on drying with gypsum formation can be found in Seck et al. (2015, 2016). We limit ourselves here to these references. A detailed comparative study of evaporation in porous media for different salts remains an open topic.

3.15. Conclusion

This overview on the evaporation of a sodium chloride salt solution in a porous medium shows that it is a very rich topic due to the couplings between evaporation, transport of ions in solution and the formation of crystallized salt structures, efflorescence or subflorescence, not to mention the links with mechanical effects, which have not been addressed at all in this chapter. Although substantial progress in understanding and modeling these situations has been made, particularly over the past decade, many open questions remain. In particular, the theory of drying with the formation of a salt crust at the surface remains to be developed. The mechanisms driving the morphological transitions of crystallized salt structures, such as the change from branched to compact structures, are not clearly understood. The original structures, such as the polygon structure on the surface of the Salar de Uyuni for example, are also poorly understood. One of the merits of the work carried out is to make it possible to clearly identify these under-researched topics and their associated challenges. In other words, the conditions are in place for a new cycle of important advances in research to occur.

However, it is important to remember that we deliberately limited ourselves to the case of NaCl and to the case of capillary-porous media. As for other transport processes, the case of nanoporous media, in particular clays, requires reconsidering the elements presented in this chapter. The fact that only NaCl solutions were considered is also a simplifying element. The case of each salt deserves to be examined specifically. Also, the presence of several salts in solution, which is not addressed in this chapter, increases the complexity and is likely to have a significant impact on evaporation.

3.16. References

Ahmad, F., Rahimi, A., Tsotsas, E., Prat, M., Kharaghani, A. (2021). From micro-scale to macro-scale modeling of solute transport in drying capillary porous media. *Int. J. Heat Mass Transfer*, 165, 120722.

Bear, J. and Bachmat, Y. (1990). *Introduction to Modeling of Transport Phenomena in Porous Media*. Springer, Berlin.

Bechtold, M., Haber-Pohlmeier, S., Vanderborght, J., Pohlmeier, A., Ferré, T.P.A., Vereecken, H. (2011). Near-surface solute redistribution during evaporation. *Geophysical Research Letters*, 38, L17404.

Bénard, H. (1901). Les tourbillons cellulaires dans une nappe liquide transportant de la chaleur par convection en régime permanent. *Annales de chimie et de physique*, 7(23), 62–144.

Bergstad, M., Or, D., Withers, P.J., Shokri, N. (2017). The influence of NaCl concentration on salt precipitation in heterogeneous porous media. *Water Resour. Res.*, 53(2), 1702–1712.

Bergstad, M., Or, D., Withers, P.J., Shokri, N. (2018). Evaporation dynamics and NaCl precipitation on capillarity-coupled heterogeneous porous surfaces. *Water Resour. Res.*, 54(6), 3876–3885.

Camassel, B., Sghaier, N., Prat, M., Ben Nasrallah, S. (2005). Ions transport during evaporation in capillary tubes of polygonal cross section. *Chem. Eng. Sci.*, 60, 815–826.

Chauvet, F., Duru, P., Geoffroy, S., Prat, M. (2009). Three periods of drying of a single square capillary tube. *Phys. Rev. Lett.*, 103, 124502.

Corti, T. and Krieger, U.K. (2007). Improved inverted bubble method for measuring small contact angles at crystal-solution-vapor interfaces. *Appl. Opt.*, 46, 5835–5839.

Dagan, G. (2012). *Flow and Transport in Porous Formations*. Springer, Berlin.

Dashtian, H., Shokri, N., Sahimi, M. (2018). Pore-network model of evaporation-induced salt precipitation in porous media: The effect of correlations and heterogeneity. *Adv. in Water Res.*, 112, 59–71.

Deegan, R.D., Bakajin, O., Dupont, T.F., Huber, G., Nagel, S.R., Witten, T.A. (1997). Capillary flow as the cause of ring stains from dried liquid drops. *Nature*, 389, 827.

Del Río, J.A. and Whitaker, S. (2016). Diffusion of charged species in liquids. *Scientific Reports*, 6, 35211.

Derluyn, H., Moonen, P., Carmeliet, J. (2014). Deformation and damage due to drying-induced salt crystallization in porous limestone. *J. Mech. Phys. Solids*, 63, 242–255.

Desarnaud, J., Derluyn, H., Carmeliet, J., Bonn, D., Shahidzadeh, N. (2014). Metastability limit for the nucleation of NaCl crystals in confinement. *J. Phys. Chem. Lett.*, 5(5), 890–895.

Desarnaud, J., Derluyn, H., Molari, L., de Miranda, S., Cnudde, V., Shahidzadeh, N. (2015). Drying of salt contaminated porous media: Effect of primary and secondary nucleation. *J. Appl. Phys.*, 118(11), 114901.

Diouf, B., Geoffroy, S., Abou-Chakra, A., Prat, M. (2018). Locus of first crystals on the evaporative surface of a vertically textured porous medium. *Eur. Phys. J. Appl. Phys.*, 81(1), 11102.

Duenas-Velasco, M. (2016). Évaporation en milieu poreux en présence de sel dissous. Structure et lois de croissance des efflorescences. Thesis, Université de Toulouse, Toulouse.

Eloukabi, H., Sghaier, N., Ben Nasrallah, S., Prat, M. (2013). Experimental study of the effect of sodium chloride on drying of porous media: The crusty-patchy efflorescence transition. *Int. J. Heat Mass Tr.*, 56, 80–93.

Grementieri, L., Molari, L., Derluyn, H., Desarnaud, J., Cnudde, V., Shahidzadeh, N., de Miranda, S. (2017). Numerical simulation of salt transport and crystallization in drying Prague sandstone using an experimentally consistent multiphase model. *Building and Environment*, 123, 289–298.

Guglielmini, L., Gontcharov, A., Aldykiewicz, A.J., Stone, H.A. (2008). Drying of salt solutions in porous materials: Intermediate-time dynamics and efflorescence. *Physics of Fluids*, 20, 077101.

Gupta, S., Huinink, H.P., Prat, M., Pel, L., Kopinga, K. (2014). Paradoxical drying due to salt crystallization. *Chem. Eng. Sci.*, 109, 204–211.

Hidri, F., Sghaier, N., Eloukabi, H., Prat, M., Ben Nasrallah, S. (2013). Porous medium coffee ring effect and other factors affecting the first crystallization time of sodium chloride at the surface of a drying porous medium. *Physics of Fluids*, 25, 127101.

Hidri, F., Diouf, B., Bouhlila, R., Geoffroy, S., Prat, M. (2019). Stagnation points as loci of solute concentration extrema at the evaporative surface of a random porous medium. *Transp. Porous Media*, 128(3), 861–879.

Huinink, H.P., Pel, L., Michels, M.A.J. (2002). How ions distribute in a drying porous medium: A simple model. *Physics of Fluids*, 14(4), 1389–1395.

Jambhekar, V.A., Helmig, R., Schröder, N., Shokri, N. (2015). Free-flow-porous-media coupling for evaporation driven transport and precipitation of salt. *Transp. Porous Media*, 110, 251–280.

Jambhekar, V.A., Mejri, E., Schröder, N., Helmig, R., Shokri, N. (2016). Kinetic approach to model reactive transport and mixed salt precipitation in a coupled free-flow-porous-media system. *Transp. Porous Media*, 114, 341–369.

Koniorczyk, M. (2012). Salt precipitation and crystallization in non-isothermal, partially saturated porous materials considering ions interaction model. *Int. J. Heat Mass Transf.*, 55, 665–679.

Koniorczyk, M. and Gawin, D. (2012). Modelling of salt crystallization in building materials with microstructure – Poromechanical approach. *Construction and Building Materials*, 36, 860–873.

Lasser, J., Nield, J.M., Goehring, L. (2020). Surface and subsurface characterization of salt pans expressing polygonal patterns. *Earth System Science Data*, 12(4), 2881–2898.

Lazhar, R., Najjari, M., Prat, M. (2020). Combined wicking and evaporation of NaCl solution with efflorescence formation: The efflorescence exclusion zone. *Physics of Fluids*, 32(6), 067106.

Le Bray, Y. and Prat, M. (1999). Three-dimensional pore network simulation of drying in capillary porous media. *Int. J. Heat Mass Tr.*, 42, 4207–4224.

Licsandru, G. (2020). Evaporative NaCl salt crusts: Formation, patterning, displacement and detachment mechanisms. PhD Thesis, Université de Toulouse, Toulouse.

Licsandru, G., Noiriel, C., Duru, P., Geoffroy, S., Abou-Chakra, A., Prat, M. (2019). Dissolution-precipitation-driven upward migration of a salt crust. *Physical Review E*, 100(3), 032802.

Licsandru, G., Noiriel, C., Geoffroy, S., Abou Chakra, A., Duru, P., Prat, M. (2022). Detachment mechanism and reduced evaporation of an evaporative NaCl salt crust. *Scientific Reports*, 12(1), 7473.

Licsandru, G., Noiriel, C., Duru, P., Geoffroy, S., Abou Chakra, A., Prat, M. (2023). Evaporative destabilization of a salt crust with branched pattern formation. *Scientific Reports*, 13(1), 1–15.

Mejri, E., Bouhlila, R., Helmig, R. (2017). Heterogeneity effects on evaporation-induced halite and gypsum coprecipitation in porous media. *Transp. Porous Media*, 118(1), 39–64.

Moghaddam, A.A., Kharaghani, A., Tsotsas, E., Prat, M. (2017). Kinematics in a slowly drying porous medium: Reconciliation of pore network simulations and continuum modeling. *Physics of Fluids*, 29(2), 022102.

Mujumdar, A.S. (2006). *Handbook of Industrial Drying*. CRC Press, Boca Raton.

Mullin, J.W. (2001). *Crystallization*, 4th edition. Butterworth Heinemann, Oxford.

Nachshon, U. and Weisbrod, N. (2015). Beyond the salt crust: On combined evaporation and subflorescent salt precipitation in porous media. *Transp. Porous Media*, 110(2), 295–310.

Nachshon, U., Weisbrod, N., Dragila, M., Grader, A. (2011a). Combined evaporation and salt precipitation in homogeneous and heterogeneous porous media. *Water Resour. Res.*, 47, W03513.

Nachshon, U., Shahraeeni, E., Or, D., Dragila, M., Weisbrod, N. (2011b). Infrared thermography of evaporative fluxes and dynamics of salt deposition on heterogeneous porous surfaces. *Water Resour. Res.*, 47, W12519.

Nachshon, U., Weisbrod, N., Katzir, R., Nasser, A. (2018). NaCl crust architecture and its impact on evaporation: Three-dimensional insights. *Geophysical Research Letters*, 45, 6100.

Naillon, A, Joseph, P., Prat, M. (2017). Sodium chloride precipitation reaction coefficient from crystallization experiment in a microfluidic device. *J. of Crystal Growth*, 463, 201–210.

Naillon, A., Joseph, P., Prat, M. (2018). Ion transport and precipitation kinetics as key aspects of the stress generation on pore walls induced by salt crystallization. *Phys. Rev. Lett.*, 120, 034502.

Naillon, A., Joseph, P., Prat, M. (2019). Direct observation of pore collapse and tensile stress generation on pore wall due to salt crystallization in a PDMS channel. *Soft Matter*, 15(22), 4562–4569.

Noiriel, C., Renard, F., Doan, M.L., Gratier, J.P. (2010). Intense fracturing and fracture sealing induced by mineral growth in porous rocks. *Chemical Geology*, 269, 197–209.

Or, D., Lehmann, P., Shahraeeni, E., Shokri, N. (2013). Advances in soil evaporation physics. A review. *Vadose Zone Journal*, 12(4), 1–16.

Pel, L., Huinink, H.P., Kopinga, K. (2002). Ion transport and crystallization in inorganic building materials as studied by nuclear magnetic resonance. *Appl. Phys. Lett.*, 81(15), 2893–2895.

Pel, L., Pishkari, R., Casti, M. (2018). A simplified model for the combined wicking and evaporation of a NaCl solution in limestone. *Materials and Structures*, 51(3), 66.

Peysson, Y. (2012). Permeability alteration induced by drying of brines in porous media. *Eur. Phys. J. Appl. Phys.*, 60(2), 24206.

Prat, M. (2002). Recent advances in pore-scale models for drying of porous media. *Chem. Eng. J.*, 86(1/2), 153–164.

Prat, M. (2007). On the influence of pore shape, contact angle and film flows on drying of capillary porous media. *Int. J. Heat Mass Tr.*, 50, 1455–1468.

Pruess, K. and Müller, N. (2009). Formation dry-out from CO_2 injection into saline aquifers 1: Effects of solids precipitation and their mitigation. *Water Resour. Res.*, 45, W03402.

Qazi, M.J., Salim, H., Doorman, C.A.W., Jambon-Puillet, E., Shahidzadeh, N. (2019). Salt creeping as a self-amplifying crystallization process. *Science Advances*, 5(12), eaax1853.

Rad, M.N., Shokri, N., Sahimi, M. (2013). Pore-scale dynamics of salt precipitation in drying porous media. *Physical Review E*, 88(3), 032404.

Rijniers, L.A., Huinink, H.P., Pel, L., Kopinga, K. (2005). Experimental evidence of crystallization pressure inside porous media. *Phys. Rev. Lett.*, 94(7), 075503.

Roy, R., Weibel, J.A., Garimella, S.V. (2022). Modeling the formation of efflorescence and subflorescence caused by salt solution evaporation from porous media. *Int. J. Heat and Mass Transfer*, 189, 122645.

Seck, M.D., Van Landeghem, M., Faure, P., Rodts, S., Combes, R., Cavalié, P., Keita, E., Coussot, P. (2015). The mechanisms of plaster drying. *Journal of Materials Science*, 50(6), 2491–2501.

Seck, M.D., Keita, E., Faure, P., Cavalié, P., Van Landeghem, M., Rodts, S., Combes, R., Coussot, P. (2016). Subflorescence and plaster drying dynamics. *Chem. Eng. Sci.*, 148, 203–211.

Sghaier, N. and Prat, M. (2009). Effect of efflorescence formation on drying kinetics of porous media. *Transport in Porous Media*, 80(3) 441–454.

Sghaier, N., Prat, M., Ben Nasrallah, S. (2006). On the influence of sodium chloride concentration on equilibrium contact angle. *Chem. Eng. J.*, 122, 47–53.

Sghaier, N., Prat, M., Ben Nasrallah, S. (2007). On ions transport during drying in a porous medium. *Transport in Porous Media*, 62, 243–274.

Sghaier, N., Geoffroy, S., Prat, M., Eloukabi, H., Ben Nasrallah, S. (2014). Evaporation driven growth of large crystallized salt structures in a porous medium. *Phys. Rev. E*, 90, 042402.

Shahidzadeh-Bonn, N. and Desarnaud, J. (2012). Damage in porous media: Role of the kinetics of salt (re) crystallization. *Eur. Phys. J. Appl. Phys.*, 60, 24205.

Shahidzadeh-Bonn, N., Desarnaud, J., Bertrand, F., Chateau, X., Bonn, D. (2010). Damage in porous media due to salt crystallization. *Physical Review E*, 81(6), 066110.

Shokri, N. (2014). Pore-scale dynamics of salt transport and distribution in drying porous media. *Physics of Fluids*, 26(1), 012106.

Shokri, N. and Or, D. (2011). What determines drying rates at the onset of diffusion-controlled stage-2 evaporation from porous media? *Water Resour. Res.*, 47(9), W09513.

Shokri-Kuehni, S.M.S., Vetter, T., Webb, C., Shokri, N. (2017a). New insights into saline water evaporation from porous media: Complex interaction between evaporation rates, precipitation, and surface temperature. *Geophysical Research Letters*, 44, 5504–5510.

Shokri-Kuehni, S.M.S., Rad, M.N., Webb, C., Shokri, N. (2017b). Impact of type of salt and ambient conditions on saline water evaporation from porous media. *Adv. in Water Res.*, 105, 154–161.

Talbi, M. and Prat M. (2019). About Schlünder's model: A numerical study of evaporation from partially wet surfaces. *Drying Technology*, 37(5), 513–524.

Talbi, M. and Prat, M. (2022). Percolating and non-percolating liquid phase continuum model of drying in capillary porous media with application to solute transport in the very low Péclet number limit. *Phys. Rev. Fluids*, 7(1), 014306.

Van Brakel, J. (1980). Mass transfer in convective drying. In *Advances in Drying*, Mujumdar, A.S. (ed.). Hemisphere, New York.

Veran-Tissoires, S. and Prat, M. (2014). Evaporation of a sodium chloride solution from a saturated porous medium with efflorescence formation. *J. Fluid Mechanics*, 749, 701–749.

Veran-Tissoires, S., Marcoux, M., Prat, M. (2012a). Discrete salt crystallization at the surface of a porous medium. *Phys. Rev. Lett.*, 108, 054502.

Veran-Tissoires, S., Marcoux, M., Prat, M. (2012b). Salt crystallization at the surface of a heterogeneous porous medium. *Europhysics Letters*, 98, 34005.

Vu, H.T. and Tsotsas, E. (2018). Mass and heat transport models for analysis of the drying process in porous media: A review and numerical implementation. *International Journal of Chemical Engineering*, 9456418.

Xu, T., Sonnenthal, E., Spycher, N., Pruess, K. (2006). TOUGHREACT – A simulation program for non-isothermal multiphase reactive geochemical transport in variably saturated geologic media: Applications to geothermal injectivity and CO_2 geological sequestration. *Computers & Geosciences*, 32(2), 145–165.

Yiotis, A.G., Salin, D., Tajerand, E.S., Yortsos, Y.C. (2012). Drying in porous media with gravity-stabilized fronts: Experimental results. *Physical Review E*, 86, 026310.

4

Poromechanics and Confined Crystallization in Porous Media

Patrick DANGLA
Navier, CNRS, École des Ponts ParisTech,
Université Gustave Eiffel, Marne-la-Vallée, France

4.1. Introduction

The crystallization of salts in porous materials is a concern for structures located by the sea, exposed to sea spray, or those located in the mountains and subject to de-icing salts, for historical monuments that are not sufficiently isolated from the ground, which are subject to rising salts by capillarity, for artistic and architectural works subject to marine environments, for archeological sites, etc. The consequences can be simply a loss of esthetic value (e.g. in the event of salt efflorescence) or more seriously the physical degradation of the material, jeopardizing the construction (case of salt subflorescence).

The degradation mechanisms caused by salts, although widely studied in the literature, are still poorly understood as there are many parameters involved. They can be grouped into three categories associated with (i) the nature of the salt, (ii) the type of material and (iii) the environmental conditions (Scherer et al. 2001; Flatt 2002; Benavente et al. 2004). The three categories, of course, interact with each other.

Salt Crystallization in Porous Media,
coordinated by Hannelore DERLUYN and Marc PRAT.

It is therefore essential to fully understand the degradation mechanisms in order to identify the most influential factors or parameters, and ultimately new methods of protecting and conserving historic structures and monuments.

In the case of subflorescence, the crystallization of salts within pores can generate high stresses in the solid matrix of the material and cause damage affecting the esthetics and resistance of the structures. The crystallization pressure predicted by Correns (1949), under certain saturation levels, is sufficient to cause damage to most building materials (Winkler and Singer 1972). Other mechanisms have also been proposed: hydration pressure (Doehne 1994), thermal expansion (Johannessen et al. 1982), climate change (Mottershead and Pye 1994), osmotic pressure (Winkler 1994) and hydraulic pressure (McMahon et al. 1992). These mechanisms are, however, controversial in the literature and none of these theories explains the phenomena better than that of crystallization pressure.

The saline solution, once deposited on the surface of the material, penetrates its pores and crystallizes within, and degrades it. The penetration of the solution is accentuated by drying and humidification processes during natural day/night or winter/summer cycles. Temperature is also a factor that can exacerbate the phenomenon by modifying the solubility constants of the salt. Important factors include the degree of (super)saturation and its location in the pore network system, crystal growth kinetics and environmental conditions (relative humidity) (Rodriguez-Navarro and Doehne 1999). On the other hand, some salts are more damaging than others (Rodriguez-Navarro and Doehne 1999).

Another important point that is not fully understood is where the salt is located within the pore network. The most popular theory is that the crystal first forms in larger pores and then progresses into the smaller ones (Wellam and Wilson 1965). This theory is, however, not completely confirmed by experiments and is sometimes contradicted (Rodriguez-Navarro and Doehne 1999). Thus, some models are based on the crystal being in the largest pores occupied by the solution (Coussy 2006). Other models consider that the salt crystals are uniformly distributed across the porous network (Derluyn et al. 2014). In addition, some salts are prone to efflorescence (halite), others to subflorescence (thenardite and mirabilite).

In the case of sodium sulfate, the conversion of thenardite into mirabilite in the presence of increasing humidity takes place by a process of dissolution and then precipitation (Rodriguez-Navarro and Doehne 1999). However, several authors claim that there is no thenardite hydration process or hydration pressure (Goudie 1977; Cooke 1981; Doehne 1994). The disparity in molar volumes between thenardite and mirabilite is not deemed the main mechanism of damage because the liquid phase must also be considered. In the case of a solution of pure water and thenardite converting to mirabilite, the system shrinks (Flatt 2002). On the other hand, the dissolution of thenardite can produce a mirabilite supersaturated solution and a significant crystallization pressure that can cause damaging stresses to develop (Flatt 2002).

Still, the crystallization pressure generated by the salt supersaturation of the interstitial solution is the most convincing theory at present to explain the development of stresses likely to cause damage to materials. In this chapter, we propose a model based on poromechanics and the crystallization pressure, allowing the prediction of macroscopic stresses and the risks of damage of the material. We will follow the approach proposed by Coussy (2006) with some improvements, namely considering the stress anisotropy within the salt crystal and the precipitation/dissolution kinetics of the salt during pore invasion and pore strain.

4.2. Chemical potential of the crystal under anisotropic stress

Phase transitions between a solution and a solid can only take place at the surface of the latter and therefore the chemical potential of solids can only be defined at the surface (Boer 1977). As several authors have pointed out (Boer 1977; Steiger 2005), the salt crystal located in the pores and in close contact with the walls of the pore will develop anisotropic and heterogeneous stresses. In doing so, the salt crystal can grow in the direction of the walls of the pore only if a film of solution is present between the wall of the pore and the solid crystal to allow the diffusive transport of the ions. This film is a nanometer thick (Scherer 1999). At the same time, the interface between the crystal and the interstitial solution, this time at the pore entrance, is subjected to a different pressure, since the crystal is in contact with the liquid. As such, the crystal is in a state of anisotropic stresses.

To simplify the analysis, the pressure exerted on the surface of the crystal in contact with the solid pore wall and that exerted at the solid–solution interface will be distinguished in what follows. This pressure, denoted by P_C, will take different values depending on the interface considered.

The chemical potential of a solid under anisotropic stress has already been proposed by Gibbs (1906), equations [393–395] then afterwards by Boer (1977) and Steiger (2005). The chemical potential of the solid only involves the local state of the solid at its surface. For elastic strain, the chemical potential of the crystal under normal stress σ_n on its surface is given by:

$$\mu_C = \psi_C - \sigma_n V_C \qquad [4.1]$$

where $\psi_C = (\psi_C)_0 + w$ is the (partial) molar free energy of the solid crystal under stress, $(\psi_C)_0$ is that when not under stress, w is the elastic molar energy and V_C is the molar volume under stress. We notice that the chemical potential of the solid depends on the orientation of its surface via the normal stress σ_n applied on the surface. However, the term $\mu_\mathrm{C} + \sigma_n V_\mathrm{C}$ that corresponds to the (partial) molar free energy of the solid crystal under stress does not depend on the surface orientation. To leading order, if the strain is small, we can consider that $w = 0$ and V_C is constant. Under these conditions, $\mu_\mathrm{C} \approx (\psi_\mathrm{C})_0 - \sigma_n V_\mathrm{C}$.

4.3. Crystallization pressure and growth kinetics

Consider a crystal of hydrated salt C composed of ions A and B under the stoichiometric form $\mathrm{C} = \mathrm{A}_{\nu_\mathrm{A}}\mathrm{B}_{\nu_\mathrm{B}}(\mathrm{H_2O})_{\nu_\mathrm{w}}$. The dissolution reaction (and the opposite precipitation reaction) is written as follows:

$$\mathrm{C} \rightleftharpoons \nu_\mathrm{A}\mathrm{A} + \nu_\mathrm{B}\mathrm{B} + \nu_\mathrm{w}\mathrm{H_2O} \quad [4.2]$$

where ν_α is the stoichiometric coefficient of α ($\alpha = \mathrm{A, B, w}$). For a hydrated crystal, $\nu_w \neq 0$. During dissolution or precipitation of the crystal, the total Gibbs energy, that is, the sum of the Gibbs energy of the crystal and the solution in contact with it, can only spontaneously decrease (at constant pressure and temperature):

$$\frac{\mathrm{d}}{\mathrm{d}t}(G) \leq 0 \quad [4.3]$$

Equilibrium is reached when G is minimum. The derivative of this Gibbs free energy is written as $\frac{\mathrm{d}}{\mathrm{d}t}(G) = \Delta_r G \frac{\mathrm{d}}{\mathrm{d}t}(\xi)$, where ξ is the reaction progress and $\Delta_r G$ is given by:

$$\Delta_r G = \sum_{\alpha=\mathrm{A,B,w}} \nu_\alpha \mu_\alpha - \mu_\mathrm{C} \quad [4.4]$$

where μ_α represents the chemical potential of α in the liquid phase (for $\alpha = \mathrm{A, B, w}$) and μ_C the chemical potential of the crystal (considered for the molecules on the crystal surface). It will subsequently be assumed that the chemical potentials of the ions and the water, μ_α, are homogeneous in the solution and the film separating the crystal from the solid wall. This means that the diffusive transportation is considered fast and that the entire liquid solution in the pore network is in thermodynamic equilibrium. The chemical potential of α is typically written as

$$\mu_\alpha = (\mu_\alpha)_0 + V_\alpha P_\mathrm{L} + RT \ln a_\alpha \quad [4.5]$$

where a_α and V_α are the activity and the partial molar volume of α in solution and P_L is the pressure of the liquid solution. If P_C is the pressure applied to the surface of the crystal, its chemical potential is written, disregarding the local elastic energy, as assumed above:

$$\mu_{\mathrm{C}} = (\psi_{\mathrm{C}})_{\mathrm{o}} + P_{\mathrm{C}} V_{\mathrm{C}} \qquad [4.6]$$

The reaction Gibbs energy then takes the form:

$$\Delta_r G = \Delta_r G^{\ominus} - V_{\mathrm{C}}(P_{\mathrm{C}} - P_0) + \bar{V}_{\mathrm{C}}(P_{\mathrm{L}} - P_0) + RT \ln a_{\mathrm{C}} \qquad [4.7]$$

where $\bar{V}_{\mathrm{C}} = \sum_{\alpha=\mathrm{A,B,w}} \nu_\alpha V_\alpha$ is the partial molar volume of the dissolved crystal and $a_{\mathrm{C}} = \prod_{\alpha=\mathrm{A,B,w}} (a_i)^{\nu_\alpha}$ the activity product of the ions for the crystal dissolution reaction. In equation [4.7], $\Delta_r G^{\ominus}$ is the standard reaction energy at T and P_0 = 1 bar which itself is related to the solubility product at temperature T and standard pressure $P_0 : \Delta_r G^{\ominus} = -RT \ln K(P_0, T)$.

Therefore:

$$\Delta_r G = -V_{\mathrm{C}}(P_{\mathrm{C}} - P_{\mathrm{L}}) + \Delta \bar{V}_{\mathrm{C}}(P_{\mathrm{L}} - P_0) + RT \ln \beta_{\mathrm{C}} \qquad [4.8]$$

where $\Delta V_{\mathrm{C}} = \bar{V}_{\mathrm{C}} - V_{\mathrm{C}}$ and $\beta_{\mathrm{C}} = a_{\mathrm{C}}/K(P_0, T)$ is the saturation index of the crystal at temperature T and standard pressure P_0.

Equation [4.8] is the generically obtained Gibbs reaction energy. As we have seen above, the pressure P_{C} and the chemical potential of the crystal μ_{C} are not homogeneous in the pore where the salt crystal is located, contrary to what was assumed for the liquid phase and μ_α. Thus, it is important to distinguish the points on the crystal surface where the reaction takes place and where the pressure P_{C} is applied , that is, the crystal–solution interface and the crystal–solid interface. We will therefore note by $P_{\mathrm{C_L}}$ the pressure exerted on the surface of the crystal in contact with the solution and $P_{\mathrm{C_S}}$ the pressure exerted on the surface of the crystal in contact with the pore wall. Pressure $P_{\mathrm{C_L}}$ is connected to the curvature, κ_{CL}, and the surface energy, γ_{CL}, of the crystal–solution interface by the relation $P_{\mathrm{C_L}} = P_{\mathrm{L}} + \gamma_{\mathrm{CL}}\kappa_{\mathrm{CL}}$. As a result, the Gibbs energy takes the following expressions.

At the crystal–solid interface:

$$\Delta_r G = -V_{\mathrm{C}}(P_{\mathrm{C_S}} - P_{\mathrm{L}}) + \Delta \bar{V}_{\mathrm{C}}(P_{\mathrm{L}} - P_0) + RT \ln \beta_{\mathrm{C}} \qquad [4.9]$$

At the crystal–solution interface:

$$\Delta_r G = -V_{\mathrm{C}}\gamma_{\mathrm{CL}}\kappa_{\mathrm{CL}} + \Delta \bar{V}_{\mathrm{C}}(P_{\mathrm{L}} - P_0) + RT \ln \beta_{\mathrm{C}} \qquad [4.10]$$

The reaction energy $\Delta_r G$ expresses the thermodynamic imbalance between the crystal and the solution. This makes it possible to consider several hypotheses for modeling the formation of the crystal. The first is the hypothesis of thermodynamic equilibrium between the crystal and the solution which imposes $\Delta_r G = 0$ everywhere. This gives the following equations:

$$P_{C_S} - P_L = \frac{\Delta \bar{V}_C}{V_C}(P_L - P_0) + \frac{RT}{V_C} \ln \beta_C \tag{4.11}$$

$$\gamma_{CL}\kappa_{CL} = \frac{\Delta \bar{V}_C}{V_C}(P_L - P_0) + \frac{RT}{V_C} \ln \beta_C \tag{4.12}$$

Equation [4.12] which expresses the equilibrium of the crystal with its solution at the crystal–solution interface is known as the Ostwald–Freundlich equation. It expresses (at $P_L = P_0$) the dependence of the solubility of the crystal on the curvature of its surface. A crystal is more soluble the smaller it is (larger curvature). Thus, large crystals will be more stable than small ones, and in the case where they are both present, the latter will disappear in favor of larger crystals. This is the Ostwald ripening phenomenon.

Equation [4.11] expresses the equilibrium of the crystal with the solution film at the crystal–solid interface. Pressure P_{C_S}, given by [4.11], is the pressure required to apply to the crystal to stop its growth. It is the crystallization pressure that Correns (1949) discovered and expressed in a simpler form.

The equilibrium hypothesis [4.11] can be justified by the fact that crystal precipitation, due to supersaturation of the film of solution, will rapidly produce the pressure necessary to stop the crystal growth, that is, for a small amount of precipitated matter. This is possible when the material is rigid enough, that is, in the elastic phase. This assumption is less obvious when the material cracks or becomes damaged because its stiffness/compliance can decrease/increase very rapidly with damage. Equation [4.11] makes it possible to predict, for example, that the crystallization pressure necessary to stop the growth of a halite crystal is approximately 68 MPa if the supersaturation of the solution is equal to 2. Such a supersaturation can very easily be reached during evaporation of water from the solution. In a spherical pore and if the porosity is not too large, the orthoradial stress at the pore wall is then approximately equal to P_{C_S}. This value exceeds the tensile strength of many materials and may explain the damages on porous materials subjected to sea spray. The equilibrium hypothesis expressed by equation [4.11] is adopted by many authors to model the mechanical behavior of poroelastic materials subjected to salt crystallization (Scherer 2004; Coussy 2006; Derluyn 2012; Grementieri et al. 2017).

Another approach is to consider a growth kinetics which respects the thermodynamic condition $\Delta_r G \frac{\mathrm{d}}{\mathrm{d}t}(\xi) \leq 0$. This is the approach adopted by many authors instead of the equilibrium hypothesis [4.12], that is, at crystal–solution interface (Derluyn 2012; Grementieri et al. 2017). A law proposed by Kirkpatrick (1975) involves linking the progress rate in proportion to $\left(1 - \exp\left(-\frac{\Delta_r G}{RT}\right)\right)$. Other authors have proposed laws of crystal growth kinetics in a more general form Nielsen and Toft (1984); Mullin (2001) expressed as a function of the saturation index β_C : $\frac{d}{dt}(\xi) = -k(\beta_C)$ with $k(\beta_C)$ a positive function for $\beta_C > 1$. In fact, it would be more exact to define a kinetic law by replacing β_C in the previous function with $\exp\left(\frac{\Delta_r G}{RT}\right)$ that is here according to [4.10]:

$$\exp\left(\frac{\Delta_r G}{RT}\right) = \frac{\beta_C}{\beta_{C_L}^{eq}} \qquad [4.13]$$

where $\beta_{C_L}^{eq}$ is the saturation index of a solution at pressure P_L in equilibrium with the crystal surface, whose curvature is κ_{CL} :

$$\ln \beta_{C_L}^{eq} = \frac{V_C}{RT}\gamma_{CL}\kappa_{CL} - \frac{\Delta \bar{V}_C}{RT}(P_L - P_0) \qquad [4.14]$$

For example by following the laws proposed by Derluyn (2012) and Grementieri et al. (2017):

$$\frac{d}{dt}(\xi) = -k_{pre}\left|\exp\left(\frac{\Delta_r G}{RT}\right) - 1\right|^{g_{pre}} \quad \text{for } \Delta_r G > 0 \qquad [4.15]$$

where k_{pre} and g_{pre} are constants to be defined (see Derluyn et al. (2014) for more information). The dissolution kinetics can be written in an equivalent form:

$$\frac{d}{dt}(\xi) = k_{dis}\left|1 - \exp\left(\frac{\Delta_r G}{RT}\right)\right|^{g_{dis}} \quad \text{for } \Delta_r G < 0 \qquad [4.16]$$

Note that these models do not take nucleation kinetics into account. To overcome this, there are models ad hoc involving supersaturation thresholds before nucleation begins (see, e.g. Derluyn et al. 2014).

Finally, we can also define growth kinetics for the crystal–solid interface in the same form as [4.15] provided that $\beta_{C_L}^{eq}$ is replaced with $\beta_{C_S}^{eq}$ in [4.13]:

$$\ln \beta_{C_S}^{eq} = \frac{V_C}{RT}(P_{C_S} - P_L) - \frac{\Delta \bar{V}_C}{RT}(P_L - P_0) \qquad [4.17]$$

which is the saturation index of a solution at pressure $P_{\rm L}$ (and temperature T) in equilibrium with the surface of the crystal under pressure $P_{\rm C_S}$ different from $P_{\rm L}$. This is the approach taken by Gu et al. (2022) to describe damage to cementitious materials due to delayed ettringite precipitation. In fact, the crystallization pressure $P_{\rm C_S}$ is no longer governed by thermodynamic equilibrium as in [4.11] but by the growth kinetics of the crystal at the crystal–solid interface and the strain coupled with the damage to the material. In the approach taken by Gu et al. (2022), the precipitation and dissolution kinetics are proportional to $\left(1 - \exp\left(-\frac{\Delta_r G}{RT}\right)\right)$, but it is obvious that another form could have been chosen.

4.4. Crystallization thermodynamics in confined media

This section is largely inspired by the article published by Coussy (2006) from which we also take the main notations. Consider an infinitesimal representative elementary volume dV_t of porous material whose porous space is occupied by a liquid phase (index J = L), a gas phase (index J = G) and a solid phase from the salt crystal (index J = C). The liquid phase is an aqueous solution composed of water (index $\alpha = \mathrm{w}$) and ions (indices $\alpha = \mathrm{A, B}$) forming the salt. The gas phase is composed of water in the vapor form (index $\alpha = \mathrm{v}$) and dry air (index $\alpha = \mathrm{a}$). Among the salt atoms in the crystal, we will make a difference between the atoms coming from the crystal–solution interface (index $\alpha = \mathrm{C_L}$) and those coming from the crystal–solid interface (index $\alpha = \mathrm{C_S}$) where necessary.

The set of indices corresponding to the constituents of the liquid, gas and crystal phases will be denoted as $\mathcal{I}(\mathrm{L})$, $\mathcal{I}(\mathrm{G})$ and $\mathcal{I}(\mathrm{C})$, respectively, that is, $\mathcal{I}(\mathrm{L}) = \mathrm{A, B, w}$, $\mathcal{I}(\mathrm{G}) = \mathrm{a, v}$ and $\mathcal{I}(\mathrm{C}) = \mathrm{C_L, C_S}$. α is the generic index of all these species that make up the three phases: $\alpha \in \mathcal{I}(\mathrm{L})$ for the liquid phase, $\alpha \in \mathcal{I}(\mathrm{G})$ for the gas phase and $\alpha \in \mathcal{I}(\mathrm{C})$ for the crystal phase.

At the scale of the continuum medium, all the constituents that we have just mentioned move in the pore space except for the salt, which is supposed to be attached to the solid skeleton. In what follows, $\mathbf{w}_\alpha$ denotes the molar flux vector of the species α.

Let us now introduce the stress tensor $\boldsymbol{\sigma}$ inside the matter contained in dV_t and the strain tensor $\boldsymbol{\epsilon}$ of the solid skeleton identified with respect to the initial configuration of the solid skeleton contained in $\mathrm{dV_0}$ at $t = 0$. Here, $F\,\mathrm{dV_0}$ is the Helmholtz free energy of the open system consisting of all the matter contained in dV_t at t. F is therefore a Lagrangian quantity of energy per unit volume of initial material. In the following, we restrict ourselves to isothermal transformations. The combination of the first and second laws of open-system thermodynamics applied to dV_t gives the inequality (see, e.g. Coussy (2010)):

$$\boldsymbol{\sigma} : \frac{\mathrm{d}}{\mathrm{d}t}(\epsilon) - \sum_{\alpha \in \mathcal{I}(\mathrm{L,G})} \mu_\alpha \nabla \cdot (\mathbf{w}_\alpha) - \frac{\mathrm{d}}{\mathrm{d}t}(F) \geq 0 \qquad [4.18]$$

In this inequality, the first two terms represent the total input of energy to the system in different forms. The first term is the rate of work done by stresses and the second term is the rate of energy brought to the system due to transported material, $-\nabla \cdot (\mathbf{w}_\alpha)$, expressed in number of moles of the constituent α. Note that only the liquid and gas phases are involved in this term. This inequality indicates that this total energy input is naturally greater than the free energy time variation of the system, the difference being the energy dissipated in the form of heat. This energy dissipation has several sources which we will now discuss.

Let $n_\alpha\, \mathrm{dV}_0$ be the number of moles of the constituent α at t in dV_t, n_α is therefore the number of moles per initial unit volume of material. The temporal variation in n_α has two sources: an external supply of matter $-\nabla \cdot (\mathbf{w}_\alpha)$ and an internal contribution due to phase changes or chemical reactions (salt precipitation-dissolution), which will be denoted $\overset{\circ}{n}_{\to\alpha}$. The equation expressing the conservation of the number of moles of the constituent α (for $\alpha = \mathrm{A, B, w, a, v, C}$) is expressed in the form:

$$\frac{\mathrm{d}}{\mathrm{d}t}(n_\alpha) = -\nabla \cdot (\mathbf{w}_\alpha) + \overset{\circ}{n}_{\to\alpha} \qquad [4.19]$$

It is clear that the internal influx of constituent α can be split into different contributions from the other constituents: $\overset{\circ}{n}_{\to\alpha} = \sum_\beta \overset{\circ}{n}_{\beta\to\alpha}$ with $\overset{\circ}{n}_{\beta\to\alpha} = -\overset{\circ}{n}_{\alpha\to\beta}$. It is easy, for example, to see that $\overset{\circ}{n}_{\to\mathrm{w}} = \overset{\circ}{n}_{\mathrm{C}\to\mathrm{w}} + \overset{\circ}{n}_{\mathrm{v}\to\mathrm{w}}$, $\overset{\circ}{n}_{\to\mathrm{a}} = 0$ and $\overset{\circ}{n}_{\to\mathrm{v}} = \overset{\circ}{n}_{\mathrm{w}\to\mathrm{v}}$. Moreover, the stoichiometry of the dissolution reaction [4.2] gives the following equations:

$$\overset{\circ}{n}_{\mathrm{C}\to\alpha} = -\nu_\alpha\, \overset{\circ}{n}_{\to\mathrm{C}} \qquad (\text{for } \alpha = \mathrm{A, B, w}) \qquad [4.20]$$

Taking into account these latter relationships and [4.19] in equality [4.18] finally gives:

$$\begin{aligned} &(\mu_\mathrm{w} - \mu_\mathrm{v})\, \overset{\circ}{n}_{\to\mathrm{v}} + \left(\sum_{\alpha \in \mathcal{I}(\mathrm{L})} \nu_\alpha \mu_\alpha \right) \overset{\circ}{n}_{\to\mathrm{C}} \\ &+ \boldsymbol{\sigma} : \frac{\mathrm{d}}{\mathrm{d}t}(\epsilon) + \sum_{\alpha = \mathcal{I}(\mathrm{L,G})} \mu_\alpha \frac{\mathrm{d}}{\mathrm{d}t}(n_\alpha) - \frac{\mathrm{d}}{\mathrm{d}t}(F) \geq 0 \end{aligned} \qquad [4.21]$$

The first term in [4.21] corresponds to the dissipation due to the liquid–vapor phase change. It is generally considered to be zero because the chemical potentials of water and steam are equal at all times: $\mu_{\mathrm{w}} = \mu_{\mathrm{v}}$. The second term in [4.21] can be rewritten to show the dissipation related to the salt dissolution–precipitation reaction, that is, $\Delta_r G \overset{\circ}{n}_{\rightarrow \mathrm{C}}$. Recognizing indeed that $\overset{\circ}{n}_{\rightarrow \mathrm{C}} = \frac{\mathrm{d}}{\mathrm{d}t}(n_{\mathrm{C}})$ is nothing but the opposite of the reaction rate introduced previously, that is, $-\frac{\mathrm{d}}{\mathrm{d}t}(\xi)$ and that this rate is split into two, that is to say $\frac{\mathrm{d}}{\mathrm{d}t}(n_{\mathrm{C_L}})$ and $\frac{\mathrm{d}}{\mathrm{d}t}(n_{\mathrm{C_S}})$, corresponding to the crystal–solution and crystal–solid interfaces where the chemical potential of the crystal takes two distinct values, denoted $\mu_{\mathrm{C_L}}$ and $\mu_{\mathrm{C_S}}$, the dissipation linked to the dissolution–precipitation reaction of the salt is written as:

$$\left(\sum_{\alpha \in \mathcal{I}(\mathrm{L})} \nu_\alpha \mu_\alpha\right) \overset{\circ}{n}_{\rightarrow \mathrm{C}} - \sum_{\alpha \in \mathcal{I}(\mathrm{C})} \mu_\alpha \frac{\mathrm{d}}{\mathrm{d}t}(n_\alpha) = \Phi_{\rightarrow} \geq 0 \qquad [4.22]$$

The left side of the inequality [4.21], corresponding to the total dissipation, is the sum of the dissipation linked to the dissolution–precipitation reaction $\Phi_{\rightarrow}$ and the intrinsic dissipation given by:

$$\boldsymbol{\sigma} : \frac{\mathrm{d}}{\mathrm{d}t}(\boldsymbol{\epsilon}) + \sum_{\alpha \in \mathcal{I}(\mathrm{L,G,C})} \mu_\alpha \frac{\mathrm{d}}{\mathrm{d}t}(n_\alpha) - \frac{\mathrm{d}}{\mathrm{d}t}(F) = \Phi_{\mathrm{int}} \geq 0 \qquad [4.23]$$

This inequality is the starting point for the formulation of the poromechanical behavior. Intrinsic dissipation is related to the irreversible behavior of the material such as viscoelasticity or plasticity. In the reversible case, the intrinsic dissipation is zero and the previous inequality becomes an equality that we will use in the next section.

4.5. State equations for poroelasticity and degrees of saturation

We consider here the case of reversible behaviors characterized by the absence of intrinsic dissipation associated with the evolution of the system. The inequality [4.23] becomes in this case an equality and as the behavior is independent of time, we obtain:

$$\boldsymbol{\sigma} : \mathrm{d}(\boldsymbol{\epsilon}) + \sum_{\alpha \in \mathcal{I}(\mathrm{L,G,C})} \mu_\alpha \mathrm{d}(n_\alpha) - \mathrm{d}(F) = 0 \qquad [4.24]$$

Let us now introduce the total porosity ϕ, and partial porosity, ϕ_{J}, porosities associated with liquid, gas and crystal phases so that $\phi\, \mathrm{dV_0}$ and $\phi_{\mathrm{J}}\, \mathrm{dV_0}$ represent the total pore volume and the J phase volume at t in dV_t. These are Lagrangian quantities. They must verify the closure relation $\phi = \sum_{\mathrm{J}} \phi_{\mathrm{J}}$.

Let us introduce the free energy $f_{\mathrm{J}}\,\mathrm{dV}_0$ of phase J contained in an infinitesimal volume dV_t, occupying, at t, the volume $\phi_{\mathrm{J}}\,\mathrm{dV}_0$. For the liquid and gas, the differential of f_{J} is written by definition:

$$\mathrm{d}\left(f_{\mathrm{J}}\right) = -P_{\mathrm{J}}\mathrm{d}\left(\phi_{\mathrm{J}}\right) + \sum_{\alpha\in\mathcal{I}(\mathrm{J})} \mu_\alpha \mathrm{d}\left(n_\alpha\right) \qquad \text{for J} = \mathrm{L}, \mathrm{G} \tag{4.25}$$

For the crystal, $\psi_\alpha = \mu_\alpha - V_{\mathrm{C}} P_\alpha$ being the partial molar free energy of the crystal on the surface α, the variation, $\mathrm{d}\left(f_{\mathrm{C}}\right)$, in free energy of the crystal is due to two contributions, that of its average molar elastic energy stored in the crystal, $\mathrm{d}\left(w_{\mathrm{C}}\right)$, and that coming from matter inputs, $\mathrm{d}\left(n_\alpha\right)$, on the surface α:

$$\mathrm{d}\left(f_{\mathrm{C}}\right) = n_{\mathrm{C}}\mathrm{d}\left(w_{\mathrm{C}}\right) + \sum_{\alpha\in\mathcal{I}(\mathrm{C})} \left(\mu_\alpha - V_{\mathrm{C}} P_\alpha\right) \mathrm{d}\left(n_\alpha\right) \tag{4.26}$$

Thereafter and for obvious simplification reasons, we assume that the variation in molar elastic energy $\mathrm{d}\left(w_{\mathrm{C}}\right)$ is negligible and that the molar volume of the crystal, V_{C}, varies very little.

Let us introduce the free energy of the skeleton by subtracting the free energy of the liquid, gas and crystal phases from F:

$$F_{\mathrm{sk}} = F - \left(\sum_{\mathrm{J=L,G,C}} f_{\mathrm{J}}\right) \tag{4.27}$$

The combination of [4.25], [4.26] (with $\mathrm{d}\left(w_{\mathrm{C}}\right) = 0$), [4.27] and [4.24] gives:

$$\mathrm{d}\left(F_{\mathrm{sk}}\right) = \boldsymbol{\sigma} : \mathrm{d}\left(\epsilon\right) + \sum_{\mathrm{J=L,G}} P_{\mathrm{J}}\mathrm{d}\left(\phi_{\mathrm{J}}\right) + \sum_{\alpha\in\mathcal{I}(\mathrm{C})} V_{\mathrm{C}} P_\alpha \mathrm{d}\left(n_\alpha\right) \tag{4.28}$$

Equation [4.28] implies that the free energy of the skeleton F_{sk} is a function of the variables $\left(\epsilon, \phi_{\mathrm{J}}, n_\alpha\right)$ (J $= \mathrm{L}, \mathrm{G}$ and $\alpha \in \mathcal{I}(\mathrm{C})$) and that the state equations are written as:

$$\boldsymbol{\sigma} = \left(\frac{\partial}{\partial \epsilon}\left(F_{\mathrm{sk}}\right)\right)_{\left(\phi_{\mathrm{J}}, n_\alpha\right)} \tag{4.29}$$

$$P_{\mathrm{J}} = \left(\frac{\partial}{\partial \phi_{\mathrm{J}}}\left(F_{\mathrm{sk}}\right)\right)_{\left(\epsilon, n_\alpha\right)} \qquad (\mathrm{J} = \mathrm{L}, \mathrm{G}) \tag{4.30}$$

$$V_{\mathrm{C}} P_\alpha = \left(\frac{\partial}{\partial n_\alpha}\left(F_{\mathrm{sk}}\right)\right)_{\left(\epsilon, \phi_{\mathrm{J}}\right)} \qquad \left(\alpha \in \mathcal{I}(\mathrm{C})\right) \tag{4.31}$$

These equations are, however, difficult to use to accurately identify the behavior law parameters using experimental data. Indeed, the energy of the skeleton F_{sk} was obtained by subtracting the energies of the fluid and solid phases from the total energy. In doing so, F_{sk} always includes the variations in energy of the interfaces between phases. These interface energies play a fundamental role in the processes of imbibition-drying with regard to the liquid and those of crystallization-dissolution with regard to the salt. In order to account for these different processes, the method followed by Coussy (2007) involves independently identifying variables related to the deformation and creation/disappearance of interfaces. For this purpose, let us break the partial porosities of each phase into two (Coussy 2010):

$$\phi_J = \phi_0 S_J + \varphi_J \qquad (J = L, G, C) \tag{4.32}$$

The first term in the right-hand side of [4.32] corresponds to the invasion process and quantifies the pore volume occupied at moment t by the fluid j before any strain occurs, that is, in the initial unstrained configuration, and the second term corresponds to the strain process and quantifies the change in the pore volume $\phi_0 S_J$ due to the effect of pressure P_J. For the special case of the crystal, $\phi_C = V_C n_C$ and $d(n_C) = d(n_{C_L}) + d(n_{C_S})$. The accretion of atoms at the crystal–solution and crystal–solid interfaces therefore contributes to the displacement of these and therefore to the increase in S_C and φ_C, respectively. A simple volume balance generated by this process makes it possible to write:

$$V_C d(n_{C_L}) = \phi_0 d(S_C), \quad V_C d(n_{C_S}) = d(\varphi_C) \tag{4.33}$$

The combination of [4.32], [4.33] and [4.28] gives:

$$d(F_{sk}) = \boldsymbol{\sigma} : d(\boldsymbol{\epsilon}) + \sum_{J=L,G,C_S} P_J d(\varphi_J) + \sum_{J=L,G,C_L} \phi_0 P_J d(S_J) \tag{4.34}$$

where for convenience, S_{C_L} is denoted by S_C and φ_{C_S} by φ_C. Equation [4.34] expresses the energy balance. The first two terms of the right-hand side correspond to the infinitesimal increment of work done by stresses and pressures associated with the strains of the solid skeleton and the volume change of pores occupied by the three phases. The last term corresponds to the increment of work associated with the displacement of the interfaces between phases with the constraint $\sum_J d(S_J) = 0$. The equality [4.34] leads to the conclusion that F_{sk} is a function of $(\epsilon, \varphi_L, \varphi_G, \varphi_C, S_L, S_C)$ with the state laws:

$$\boldsymbol{\sigma} = \left(\frac{\partial}{\partial \epsilon} (F_{\text{sk}}) \right)_{(\varphi_\text{J}, S_\text{L}, S_\text{C})} \quad [4.35]$$

$$P_\text{J} = \left(\frac{\partial}{\partial \varphi_\text{J}} (F_{\text{sk}}) \right)_{(\epsilon, S_\text{L}, S_\text{C})} \qquad (\text{J} = \text{L}, \text{G}, \text{C}_\text{S}) \quad [4.36]$$

$$\phi_0 (P_\text{G} - P_\text{L}) = - \left(\frac{\partial}{\partial S_\text{L}} (F_{\text{sk}}) \right)_{(\epsilon, \varphi_\text{J})} \quad [4.37]$$

$$\phi_0 (P_{\text{C}_\text{L}} - P_\text{G}) = \left(\frac{\partial}{\partial S_\text{C}} (F_{\text{sk}}) \right)_{(\epsilon, \varphi_\text{J})} \quad [4.38]$$

It is useful to notice that the passage of the state equations [4.29-4.31] to [4.35-4.38] results in an additional state equation. This is due to the decomposition [4.32] for liquid and gas, which introduces an additional state variable (given the closure relation $\sum_\text{J} S_\text{J} = 1$). In this new set of state variables, these must still be considered independent. To fully understand why, it is necessary to consider the thermodynamic force, $\phi_0(P_\text{G} - P_\text{L})$, associated with S_L in the state equation [4.37] as having a different status from those of P_L and P_G in the equations [4.36] even if, *in fine*, their numerical values are related. In fact, the force $\phi_0(P_\text{G} - P_\text{L})$ is associated with the invasion of the pores by the liquid, that is to say with the energy variations of the interfaces while the forces P_J are associated with the volume change of the pores involving the energy stored in the solid skeleton. To clearly distinguish between these two types of energy, we will break F_{sk} into two parts: an elastic energy associated with the sole solid skeleton, W, and a surface energy (per unit pore volume) associated with phase interfaces, U. For the sake of simplification, we will assume that the energy of the interfaces, U, does not vary significantly with strain variables, $(\epsilon, \varphi_\text{J})$, so that U will only depend on S_L and S_C (see, e.g. Dangla and Pereira (2014) for taking into account the strain in U):

$$F_{\text{sk}} = W(\epsilon, \varphi_\text{L}, \varphi_\text{G}, \varphi_\text{C}, S_\text{L}, S_\text{C}) + \phi_0 U(S_\text{L}, S_\text{C}) \quad [4.39]$$

Substituting [4.39] in [4.35-4.38] gives:

$$\boldsymbol{\sigma} = \left(\frac{\partial}{\partial \epsilon} (W) \right)_{(\varphi_\text{J}, S_\text{L}, S_\text{C})} \quad [4.40]$$

$$P_\text{J} = \left(\frac{\partial}{\partial \varphi_\text{J}} (W) \right)_{(\epsilon, S_\text{L}, S_\text{C})} \qquad (\text{J} = \text{L}, \text{G}, \text{C}_\text{S}) \quad [4.41]$$

$$P_{\mathrm{G}} - P_{\mathrm{L}} = -\left(\frac{\partial}{\partial S_{\mathrm{L}}}(W/\phi_0 + U)\right)_{(\epsilon,\varphi_{\mathrm{J}},S_{\mathrm{C}})} \quad [4.42]$$

$$P_{\mathrm{C_L}} - P_{\mathrm{G}} = \left(\frac{\partial}{\partial S_{\mathrm{C}}}(W/\phi_0 + U)\right)_{(\epsilon,\varphi_{\mathrm{J}},S_{\mathrm{L}})} \quad [4.43]$$

In linear poroelasticity, the energy W of the porous material is a quadratic form of the variables $(\epsilon, \varphi_{\mathrm{L}}, \varphi_{\mathrm{G}}, \varphi_{\mathrm{C}})$ whose coefficients, as a factor of these variables, are related to the elastic properties of the material. These coefficients depend on S_{L} and S_{C} as indicated in W. As a result, the derivatives $\frac{\partial}{\partial S_{\mathrm{J}}}(W)$ are also quadratic forms obtained by replacing the coefficients of the quadratic form of W by their derivatives with respect to S_{J}. They are therefore second-order quantities with respect to $(\epsilon, \varphi_{\mathrm{L}}, \varphi_{\mathrm{G}}, \varphi_{\mathrm{C}})$, that is, involving terms such as $\epsilon_{ij}\epsilon_{kl}$, $\epsilon_{ij}\varphi_{\mathrm{J}}$, $\varphi_{\mathrm{J}}\varphi_{\mathrm{K}}$. Consequently, assuming small strain, one can neglect the terms $\frac{\partial}{\partial S_{\mathrm{J}}}(W/\phi_0)$ in [4.42-4.43] and so we get:

$$P_{\mathrm{G}} - P_{\mathrm{L}} = -\left(\frac{\partial}{\partial S_{\mathrm{L}}}(U)\right)_{(S_{\mathrm{C}})} \quad [4.44]$$

$$P_{\mathrm{C_L}} - P_{\mathrm{G}} = \left(\frac{\partial}{\partial S_{\mathrm{C}}}(U)\right)_{(S_{\mathrm{L}})} \quad [4.45]$$

such that:

$$\mathrm{d}(U) = -(P_{\mathrm{G}} - P_{\mathrm{L}})\mathrm{d}(S_{\mathrm{L}}) + (P_{\mathrm{C_L}} - P_{\mathrm{G}})\mathrm{d}(S_{\mathrm{C}}) \quad [4.46]$$

We assumed that U did not depend on the pore strain φ_{J}. Otherwise, the state equations [4.41] would have introduced an additional term on the left-hand side, depending on the derivative of U with respect to φ_{J}. This term can be interpreted as the effect of surface energies on the pressure transmitted to the solid skeleton, an effect similar to Laplace's equation (see, e.g. Coussy (2010); Dangla and Pereira (2014) for an explicit consideration of these terms). The fact that U does not depend on φ_{J} could, in principle, be verified experimentally by showing that U, obtained by integrating [4.46], is insensitive to the intensity of pressures P_{J} (Coussy 2006).

Let us introduce the Legendre–Fenchel transform W^* of W in the form:

$$W^* = W - \sum_{\mathrm{J=L,G,C_S}} P_{\mathrm{J}}\varphi_{\mathrm{J}} \quad [4.47]$$

This is a function of $(\epsilon, P_{\text{L}}, P_{\text{G}}, P_{\text{C}_\text{S}}, S_{\text{L}}, S_{\text{C}})$. The state equations [4.40-4.41] are then inverted in the form:

$$\boldsymbol{\sigma} = \left(\frac{\partial}{\partial \boldsymbol{\epsilon}}(W^*)\right)_{(P_{\text{J}}, S_{\text{L}}, S_{\text{C}})} \qquad [4.48]$$

$$\varphi_{\text{J}} = -\left(\frac{\partial}{\partial P_{\text{J}}}(W^*)\right)_{(\boldsymbol{\epsilon}, S_{\text{L}}, S_{\text{C}})} \qquad (\text{J} = \text{L}, \text{G}, \text{C}_\text{S}) \qquad [4.49]$$

These equations are the general poroelasticity equations. As indicated above, linear poroelasticity involves adopting a quadratic form of the variables of W^*, that is, $(\boldsymbol{\epsilon}, P_{\text{J}})$. Moreover, in the case of isotropic materials, the state equations are written as:

$$\sigma_{ij} = (K - 2G/3)\epsilon\delta_{ij} + 2G\epsilon_{ij} - \sum_{\text{J}=\text{L},\text{G},\text{C}_\text{S}} b_{\text{J}} P_{\text{J}} \delta_{ij} \qquad [4.50]$$

$$\varphi_{\text{J}} = b_{\text{J}}\epsilon + \sum_{\text{K}=\text{L},\text{G},\text{C}_\text{S}} P_{\text{K}}/N_{\text{JK}} \qquad [4.51]$$

where $\epsilon = \text{tr}(\boldsymbol{\epsilon})$. In these equations, K and G are, respectively, the compression modulus and the shear modulus of the drained material, that is to say subjected to zero pressures within the pores, $P_{\text{J}} = 0$. Consequently, K and G do not depend on the degrees of saturation S_{J}. Coefficients b_{J} and N_{JK} are, respectively, the generalized Biot coefficient and Biot modulus. They verify Maxwell's symmetry conditions $N_{\text{JK}} = N_{\text{KJ}}$. Unlike K and G, these coefficients generally depend on the degrees of saturation S_{J}.

4.6. Identification of poroelastic properties

Determining the poroelastic coefficients requires macroscopic experimental data, which are often difficult to obtain. Upscaling homogenization methods are one of the ways to estimate some of these coefficients from the microstructure of the material and the microscopic mechanical properties of the matrix constituting the solid skeleton. This is the subject of this section.

The first relations are obtained by studying the response of the material subjected to an isotropic confinement of its solid skeleton, that is, $\sigma_{ij} = -P\delta_{ij}$ and $P_{\text{J}} = P$. Assuming that the solid skeleton is elastic, homogeneous and isotropic at the microstructure scale, the relative variation in its volume (as well as that of the pores) will simply be given by $-P/k_s$, where k_s is the compression modulus of the matrix. By using these relations in [4.50-4.51], it immediately becomes (see Coussy (2004)):

$$b = \sum_{\mathrm{J=L,G,C}} b_\mathrm{J} = 1 - K/k_s, \qquad \sum_{\mathrm{K=L,G,C}} \frac{1}{N_\mathrm{JK}} = \frac{b_\mathrm{J} - \phi_0 S_\mathrm{J}}{k_s} \tag{4.52}$$

where, for convenience: $b_\mathrm{C} = b_{\mathrm{C_S}}$ and $N_\mathrm{JC} = N_{\mathrm{JC_S}}$. These equations are general. They ensure the compatibility of the behavior equations at the microscopic and macroscopic scales under the condition of the homogeneity of the elastic matrix constituting the skeleton. The first relation of [4.52] shows that the coefficients b_J, functions of S_L and S_C, are not independent (b being a constant) and that it allows the introduction of functions, $\chi_\mathrm{J}(S_\mathrm{L}, S_\mathrm{C})$, of the same type as Bishop's coefficient, such that:

$$b_\mathrm{J} = b\chi_\mathrm{J}(S_\mathrm{L}, S_\mathrm{C}) \tag{4.53}$$

and satisfying the condition $\sum_\mathrm{J} \chi_\mathrm{J} = 1$. Determining these Bishop coefficients requires additional information through homogenization methods taking into account the multi-scale nature of the pore network (Coussy 2006; Coussy and Brisard 2009). A stronger hypothesis, very often made implicitly or explicitly, involves assuming the isodeformation of the pores occupied by the three phases at zero pressure:

$$\frac{\varphi_\mathrm{L}}{\phi_0 S_\mathrm{L}} = \frac{\varphi_\mathrm{G}}{\phi_0 S_\mathrm{G}} = \frac{\varphi_\mathrm{C}}{\phi_0 S_\mathrm{C}} \quad \text{for } P_\mathrm{L} = P_\mathrm{G} = P_{\mathrm{C_S}} = 0 \tag{4.54}$$

leading to the identification of:

$$b_\mathrm{J} = bS_\mathrm{J}\,, \quad \sum_{\mathrm{K=L,G,C}} \frac{1}{N_\mathrm{JK}} = \frac{S_\mathrm{J}}{N}\,, \ \frac{1}{N} = \frac{b - \phi_0}{k_s} \tag{4.55}$$

This isodeformation hypothesis is also compatible with the standard Mori-Tanaka scheme Mori and Tanaka (1973) and the self-consistent scheme as a family of spherical cavities included in an elastic medium either with the same elastic properties as the matrix of the actual material (Mori-Tanaka diagram: $k = k_s$ and $g = g_s$) or with the same elastic properties as the desired equivalent homogeneous medium (self-consistent scheme: $k = K$ and $g = G$). For information purposes, the results of these approaches are given below (Dormieux et al. 2006; Coussy and Brisard 2009):

$$K = (1 - \phi_0) k_s \frac{4g}{3\phi_0 k_s + 4g} \tag{4.56}$$

$$G = (1 - \phi_0) g_s \frac{9k + 8g}{9k(1 + \frac{2}{3}\phi_0 g_s/g) + 8g(1 + \frac{3}{2}\phi_0 g_s/g)} \quad [4.57]$$

$$\frac{1}{N_{\text{JK}}} = \frac{b - \phi_0}{K} \left(\sum_{\text{I}} (S_{\text{I}} \delta_{\text{JI}} \delta_{\text{KI}}) - b S_{\text{J}} S_{\text{K}} \right) \quad [4.58]$$

The Mori-Tanaka approach involves choosing $k = k_s$, $g = g_s$ while the self-consistent approach involves choosing $k = K$ and $g = G$. In the latter case, the modules K and G are given implicitly by the expressions above (see, e.g. Coussy and Brisard (2009) for the solution). For more information on homogenization methods, the reader is invited to consult the work by Dormieux et al. (2006).

Approaches that go beyond the isodeformation hypothesis require the use of scale separation hypotheses between pore families (see, e.g. Coussy 2006; Coussy and Brisard 2009).

4.7. Water retention curve and distribution of salt crystals

We will now use the state laws [4.44] and [4.45]. Let us first examine the situations where the material pores are filled only by the liquid and gas phases. We are in the presence of a partially saturated material, the subject of numerous studies. The state law [4.44] linking the degree of liquid saturation and capillary pressure $P_{\text{cap}} = P_{\text{G}} - P_{\text{L}}$ is expressed by a water retention curve in the form:

$$S_{\text{L}} = \mathcal{S}_{\text{P}}(P_{\text{cap}}) \quad [4.59]$$

As Coussy (2006) indicated, this equation is the macroscopic translation of Laplace's equation linking, at the microscopic scale, the capillary pressure, the surface energy γ_{LG} and the curvature of the liquid–gas interface κ_{LG}: $P_{\text{cap}} = \gamma_{\text{LG}} \kappa_{\text{LG}}$. The curvature of the interface can be related to the pore entry radius by $\kappa_{\text{LG}} = 2\cos(\theta_{\text{LG}})/r_{\text{LG}}$ (θ_{LG} is the contact angle of the liquid–gas interface with the solid wall) so that a relationship can be established between the degree of saturation and the pore entry radius where the liquid–gas interface is located, that is, the largest pore occupied by the liquid: $\mathcal{S}_{\text{P}}(P_{\text{cap}}) = \mathcal{S}_{\text{r}}(r_{\text{LG}})$. The function $\mathcal{S}_{\text{r}}(r)$ characterizes the pore size distribution of the material and can be measured by mercury porosimetry with the necessary precautions (Daïan 2014). The function $\mathcal{S}_{\text{P}}(P_{\text{cap}})$ can then be theoretically identified with $\mathcal{S}_{\text{r}}(2\gamma_{\text{LG}} \cos(\theta_{\text{LG}})/P_{\text{cap}})$.

Let us subject the material to isothermal drying caused by a drop in relative humidity h_r. If the solution contains dissolved salt, this will lower the chemical activity of the water a_{w} below 1 and the capillary pressure is then given by Kelvin's law:

$$P_{\mathrm{G}} - P_{\mathrm{L}} = -\frac{RT}{V_{\mathrm{w}}} \ln\left(\frac{h_r}{a_{\mathrm{w}}}\right) \qquad [4.60]$$

The combination of Laplace's law and Kelvin's law gives:

$$\frac{2}{r_{\mathrm{LG}}} = -\frac{RT}{\gamma_{\mathrm{LG}} \cos(\theta_{\mathrm{LG}}) V_{\mathrm{w}}} \ln\left(\frac{h_r}{a_{\mathrm{w}}}\right) \qquad [4.61]$$

The Kelvin–Laplace law [4.61] tells us the value of the largest pore entry radius occupied by the liquid solution.

In the presence of crystallized salts, that is, when $S_{\mathrm{C}} > 0$, how can we generalize the law [4.59] to identify the state equation [4.44] ? The answer to this question is not unanimous in the literature. We examine the two most used approaches.

4.7.1. *Thermodynamic equilibrium for the crystal–solution interface*

The choice adopted by Coussy (2006) is based on the assumption that the drop in relative humidity is slow enough for the salt concentration in the solution to be homogeneous at all times and for the solution to be in thermodynamic equilibrium with the crystal, that is, [4.12] is verified:

$$\gamma_{\mathrm{CL}} \kappa_{\mathrm{CL}} = \frac{\Delta \bar{V}_{\mathrm{C}}}{V_{\mathrm{C}}} (P_{\mathrm{L}} - P_0) + \frac{RT}{V_{\mathrm{C}}} \ln \beta_{\mathrm{C}} \qquad [4.62]$$

Assuming the gas pressure is at atmospheric pressure P_0, we obtain by using Kelvin's law [4.60] and connecting the curvature κ_{CL} to the pore entrance radius and the crystal–liquid–solid contact angle, $\kappa_{\mathrm{CL}} = 2\cos(\theta_{\mathrm{CL}})/r_{\mathrm{CL}}$:

$$\frac{2}{r_{\mathrm{CL}}} = \frac{RT}{\gamma_{\mathrm{CL}} \cos(\theta_{\mathrm{CL}}) V_{\mathrm{C}}} \left(\frac{\Delta \bar{V}_{\mathrm{C}}}{V_{\mathrm{w}}} \ln\left(\frac{h_r}{a_{\mathrm{w}}}\right) + \ln \beta_{\mathrm{C}}\right) \qquad [4.63]$$

This equation gives the value of the pore entry radius where the crystal–solution interface is located. For a fixed humidity, the right-hand side is a function of the activities a_α themselves depending on the salt concentration. The activity of the ions and that of the water are, respectively, increasing and decreasing functions of the salt concentration. However, for common salt concentration values, the saturation index,

β_C, is an increasing function of concentration. The right side of [4.63] is therefore an increasing function of the salt concentration, so that following the increase in the salt concentration, by evaporation or by external salt input, the crystallization will first appear in the pores, saturated in solution, whose entry radius is the largest. If the salt concentration increases further, the salt crystal will then penetrate the smaller and smaller pores respecting thermodynamic equilibrium [4.63]. The analysis proposed by Coussy states that the largest of the pore entry radii occupied by the solution is equal to the smallest of the pore entry radii occupied by the salt crystal. Under these conditions the pore entrance radii r_{LG} and r_{CL}: $r_{LG} = r_{CL}$. Consequently, the law [4.59] remains valid in the presence of salts while the salt and gas occupy the space left by the liquid: $S_C + S_G = 1 - S_L$. Finally, the equality of relations [4.61] and [4.63] links the crystal saturation index, β_C, water activity, and the humidity:

$$\ln \beta_C = -\left(\frac{\gamma_{CL}\cos(\theta_{CL})}{\gamma_{LG}\cos(\theta_{LG})} + \frac{\Delta \bar{V}_C}{V_C}\right)\frac{V_C}{V_w}\ln\left(\frac{h_r}{a_w}\right) \quad [4.64]$$

This latter equation implicitly relates salt concentration to moisture. The crystallization pressure is given by:

$$P_{C_S} - P_L = \left(\frac{\gamma_{CL}\cos(\theta_{CL})}{\gamma_{LG}\cos(\theta_{LG})}\right)(P_G - P_L) \quad [4.65]$$

The hypothesis $r_{LG} = r_{CL}$, formulated by Coussy, is illustrated in Figure 4.1 taken from Coussy (2006) in the case of the thermodynamic equilibrium of mirabilite ($C = Na_2SO_4.10H_2O$) with its solution at different relative humidities as expressed by the relation [4.64] where β_C is almost x/x_0 with x the molar fraction of dissolved mirabilite and x_0 the solubility of mirabilite at 20řC and 1 bar ($x_0 = 0.02235$).

4.7.2. *Thermodynamic out-of-equilibrium for the crystal–solution interface*

This is the approach adopted in some studies (Derluyn et al. 2014; Grementieri et al. 2017). The next chapter will give more complete illustrations of the models resulting from this approach.

The evolution of the degree of crystal saturation is governed by kinetics similar to [4.15] for precipitation and [4.16] for dissolution:

$$\phi_0 \frac{\mathrm{d}}{\mathrm{d}t}(S_C) = k_{pre}\left|\exp\left(\frac{\Delta_r G}{RT}\right) - 1\right|^{g_{pre}} \quad \text{for } \Delta_r G > 0 \quad [4.66]$$

$$\phi_0 \frac{\mathrm{d}}{\mathrm{d}t}(S_C) = -k_{dis}\left|1 - \exp\left(\frac{\Delta_r G}{RT}\right)\right|^{g_{dis}} \quad \text{for } \Delta_r G < 0 \quad [4.67]$$

where the constants k and g are to be calibrated according to the nature of the crystal (see Derluyn (2012)). Salt is assumed to be evenly distributed in the pore network without modifying the pore size distribution so that, at any time, the pore volume available for the liquid and the gas is given by $\phi_0 - \phi_C = \phi_0(1 - S_C)$. The degree of saturation of the liquid is then simply decreased by a factor equal to $(1 - S_C)$ and given by $S_L = (1 - S_C)\mathcal{S}_P(P_{cap})$. In these equations, the Gibbs reaction energy, $\Delta_r G$, is given by equation [4.10] in which $\kappa_{CL} = 0$. The crystallization pressure is given by equation [4.11].

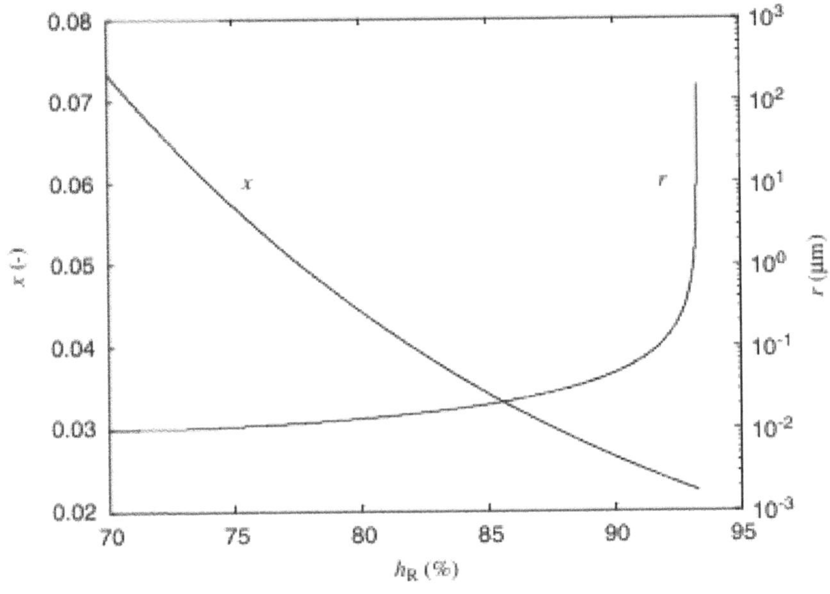

Figure 4.1. *Molar fraction x and pore entry radius $r = r_{LG} = r_{CL}$ corresponding to the equilibrium of the mirabilite with its solution at different relative humidities (Coussy 2006)*

4.8. Conclusion

This chapter presented a poromechanical model for porous materials partially saturated by a liquid solution, a wet gas and a solid formed by salt crystals. The poroelastic properties were identified by upscaling techniques within the framework of the pore isodeformation hypothesis, for which the Biot coefficients are proportional to the degrees of saturation of each phase. This last point is undoubtedly a way of improving the model, which requires the application of an upscaling method at several levels taking into account the multi-scale nature of the pores occupied by the various phases.

In this model, the salt crystal is in a state of anisotropic stresses, that is, it exerts a pressure on the walls of the solid skeleton different from the pressure due to contact with the fluid phase. This model makes it possible to consider salt growth kinetics at the crystal–solution and crystal–solid interfaces. For example, some models in the literature consider kinetics associated with the growth of salt crystals at the crystal–solution interface but consider a thermodynamic equilibrium at the crystal–solid interface. Others consider that the crystal is constantly in thermodynamic equilibrium with the solution. This model makes it possible to consider the two approaches in the same framework.

Modeling the water retention curve in the presence of salt crystals and their distribution in the pore network is presented in the literature according to two very different approaches. We present these two approaches in this same modeling framework and we show that these two approaches are somehow associated with the modeling choices concerning the precipitation–dissolution of the salt crystal. A first model, based on thermodynamic equilibrium, shows that the salt crystallizes in the largest pores occupied by the liquid. Another model, based on precipitation kinetics, assumes that the salt is evenly distributed in the pore network without changing the pore size distribution. Research is still needed to determine which parameters govern the distribution of salt crystals in the pores. How can we also explain that certain salts, such as sodium chloride, tend to form a type of crystallization, which preferentially takes place on the surface of the material without causing damage (efflorescence) while others, such as mirabilite, take place inside (subflorescence)? Coussy, for example, attributes this phenomenon to variations in liquid–gas surface tension depending on the nature of the dissolved salt. For example, the surface tension of a sodium chloride solution would be greater than that of a sodium sulfate solution, which would induce greater capillary pressure gradients (and therefore fluxes) in the case of sodium chloride. It would undoubtedly be necessary to continue research in order to confirm or invalidate this hypothesis by experimental and numerical means.

Finally, the model presented here is purely elastic. In the case of certain materials where the failure is relatively ductile, it would be useful to extend this model to irreversible behaviors such as plasticity or damage.

4.9. References

Benavente, D., del Cura, M., Garcia-Guinea, J., Sánchez-Moral, S., Ordonez, S. (2004). Role of pore structure in salt crystallisation in unsaturated porous stone. *Journal of Crystal Growth*, 260(3), 532–544.

Boer, R. (1977). On the thermodynamics of pressure solution – Interaction between chemical and mechanical forces. *Geochimica et Cosmochimica Acta*, 41, 249–256.

Cooke, R.U. (1981). Salt weathering in deserts. *Proceedings of the Geologist Association*, 92, 1–16.

Correns, C. (1949). Growth and dissolution of crystals under linear pressure. *Discussions of the Faraday Society*, 5, 267–271.

Coussy, O. (2004). *Poromechanics*. Wiley, New York.

Coussy, O. (2006). Deformation and stress from in-pore drying-induced crystallization of salt. *Journal of the Mechanics and Physics of Solids*, 54, 1517–1547.

Coussy, O. (2007). Revisiting the constitutive equations of unsaturated porous solids using a Lagrangian saturation concept. *International Journal for Numerical and Analytical Methods in Geomechanics*, 31(15), 1675–1694.

Coussy, O. (2010). *Mechanics and Physics of Porous Solids*. Wiley, New York.

Coussy, O. and Brisard, S. (2009). Prediction of drying shrinkage beyond the pore isodeformation assumption. *Journal of Mechanics of Materials and Structures*, 4(2), 263–279.

Daïan, J. (2014). *Equilibrium and Transfer in Porous Media 1: Equilibrium States*. ISTE Ltd, London, and John Wiley & Sons, New York.

Dangla, P. and Pereira, J. (2014). A thermodynamical approach of effective stresses in unsaturated soils incorporating the concept of partial pore deformations. *Vadose Zone Journal*, 13(5), 1–11.

Derluyn, H. (2012). Salt transport and crystallization in porous limestone – X-ray imaging and poromechanical modeling. Thesis, ETH, Zurich.

Derluyn, H., Moonen, P., Carmeliet, J. (2014). Deformation and damage due to drying-induced salt crystallization in porous limestone. *Journal of the Mechanics and Physics of Solids*, 63, 242–255.

Doehne, E. (1994). In situ dynamics of sodium sulfate hydration and dehydration in stone pores: Observation at high magnification using environmental scanning microscopy. In *Third International Congress on the Conservation of Monuments in the Mediterranean Basin*, Fassina, V., Ott, H., Zezza, F. (eds).

Dormieux, L., Kondo, D., Ulm, F. (2006). *Microporomechanics*. Wiley, New York.

Flatt, R. (2002). Salt damage in porous materials: How high supersaturations are generated. *Journal of Crystal Growth*, 242, 435–454.

Gibbs, J.W. (1906). *The Scientific Papers of J. Willard Gibbs, Vol. I. Thermodynamics*. Longmans/Green and Co, London/New York/Bombay.

Goudie, A.S. (1977). Sodium sulphate weathering and the disintegration of Mohenjo-daro, Pakistan. *Earth Surface Processes*, 2(1), 75–86.

Grementieri, L., Molari, L., Derluyn, H., Desarnaud, J., Cnudde, V., Shahidzadeh, N., de Miranda, S. (2017). Numerical simulation of salt transport and crystallization in drying Prague sandstone using an experimentally consistent multiphase model. *Building and Environment*, 123, 289–298.

Gu, Y., Dangla, P., Martin, R.-P., Omikrine Metalssi, O., Fen-Chong, T. (2022). Modeling the sulfate attack induced expansion of cementitious materials based on interface-controlled crystal growth mechanisms. *Cement and Concrete Research*, 152, 106676.

Johannessen, C.L., Freiereisen, J.J., Wells, A.N. (1982). Weathering of ocean cliffs by salt expansion in a mid-latitude coastal environment. *Shore and Beach*, 50, 26–34.

Kirkpatrick, R. (1975). Crystal growth from the melt: A review. *American Mineralogist*, 60, 798–814.

McMahon, D.J., Sandberg, P., Folliard, K., Mehta, P.K. (1992). Deterioration mechanisms of sodium sulfate. In *VII International Congress on Deterioration and Conservation of Stone*, volume 2. LNEC, Lisbon.

Mori, K. and Tanaka, K. (1973). Average stress in matrix and average elastic energy of materials with misfitting inclusions. *Acta Metallurgica*, 21(5), 1605–1609.

Mottershead, D.N. and Pye, K. (1994). Tafoni on coastal slopes, South Devon, U.K. *Earth Surface Processes and Landforms*, 19, 543–563.

Mullin, J. (2001). *Crystallization*, 4th edtion. Butterworth-Heinemann, Oxford.

Nielsen, A. and Toft, J. (1984). Electrolyte crystal growth kinetics. *Journal of Crystal Growth*, 67(2), 278–288.

Rodriguez-Navarro, C. and Doehne, E. (1999). Salt weathering: Influence of evaporation rate, supersaturation and crystallization pattern. *Earth Surface Processes and Landforms*, 24, 191–209.

Scherer, G. (1999). Crystallization in pores. *Cement and Concrete Research*, 29, 1347–1358.

Scherer, G. (2004). Stress from crystallization of salt. *Cement and Concrete Research*, 34, 1613–1624.

Scherer, G., Flatt, R., Wheeler, G. (2001). Materials science research for the conservation of sculpture and monuments. *MRS Bulletin*, 26(1), 44–50.

Steiger, M. (2005). Crystal growth in porous materials I: The crystallization pressure of large crystals. *Journal of Crystal Growth*, 282, 455–469.

Wellam, H. and Wilson, A. (1965). Salt weathering, a neglected geological erosive agent in coastal and arid environments. *Nature*, 205, 1097–1098

Winkler, E.M. (1994). *Stone in Architecture*. Springer, Berlin.

Winkler, E. and Singer, P. (1972). Crystallization pressure of salt in stone and concrete. *Geological Society of America Bulletin*, 83, 3509–3514.

5

Experimental Observations on Salt Crystallization in Geomaterials

Hannelore DERLUYN[1,2]

[1] *LFCR, CNRS, Université de Pau et des Pays de l'Adour, France*

[2] *DMEX, CNRS, Université de Pau et des Pays de l'Adour, France*

5.1. Introduction

The previous chapters have introduced fundamental concepts related to salt crystallization in porous media. This chapter focuses on an experimental assessment of these concepts by looking inside geomaterials while salt transport, salt crystallization and crystallization-induced damage occur. To that extent, non-destructive techniques are employed, such as NMR (e.g. Pel et al. 2018) or X-ray and neutron imaging (e.g. Cnudde et al. 2015). The obtained datasets provide a richness for understanding the coupled effects of salt crystallization in porous media. In addition, they serve for the validation of numerical models that are developed by engineers to predict salt precipitation and salt damage risks in construction materials or cultural heritage.

The chapter will focus on the use of non-destructive imaging techniques, that is, X-ray and neutron imaging, for the experimental study of geomaterials. The chapter starts with a short introduction on X-ray and neutron imaging, and then continues with two illustrative examples to elucidate on salt crystallization in a limestone and a sandstone. The focus lies on NaCl crystallization, but the techniques can be used for any kind of salt precipitation in geomaterials. The experimental characterization

Salt Crystallization in Porous Media,
coordinated by Hannelore DERLUYN and Marc PRAT.

and quantification are discussed in detail, and the link is made with numerical simulation studies in which the obtained datasets were used.

5.2. X-ray and neutron imaging

This section introduces X-ray and neutron imaging, the two experimental techniques that are used in this chapter to obtain data on salt transport, salt crystallization and crystallization damage in geomaterials. Transmission-based imaging techniques have been described in the literature as an ideal tool to investigate the structure of geomaterials and dynamic processes (transport, crystallization) occurring therein (Cnudde and Boone 2013; Wildenschild and Sheppard 2013; Maire and Withers 2014; Perfect et al. 2014). The advantage of these techniques is that they are non-destructive. In the first part of this section, the interaction of X-rays and neutrons with matter is described, as well as their complementary nature. In the second part, the image acquisition process is discussed.

5.2.1. *Interaction of X-rays and neutrons with matter*

Three types of interaction of radiation with matter can be distinguished:

– Transmission: the radiation goes through the material without changing energy or direction.

– Absorption: the radiation is absorbed transferring its energy to the absorbing material.

– Scattering: the direction of the radiation changes, and also the energy can be lowered.

The difference between X-ray and neutron radiation lies in the interaction probability of the respective radiation with the different constituting elements of the material. A measure for this interaction probability is given by the attenuation coefficient (macroscopic cross-section) μ. The intensity of the attenuated beam I is then described by the Lambert–Beer law:

$$I = I_0 e^{-\mu d} \quad [5.1]$$

where I_0 is the intensity of the incident beam and d the thickness of the object. The attenuation coefficient is given by the product of the microscopic cross-section σ_m and the atomic density N_{atom} (number of atoms per volume unit):

$$\mu = \sigma_m N_{atom} \quad [5.2]$$

The microscopic cross-section σ_m represents the probability for an X-ray photon or a neutron, traveling with a certain velocity, to interact with a single atom, and sums up the probabilities of the different types of interaction. The probability for every type of interaction depends on the energy of the X-ray photon or the neutron and on the specific element it crosses.

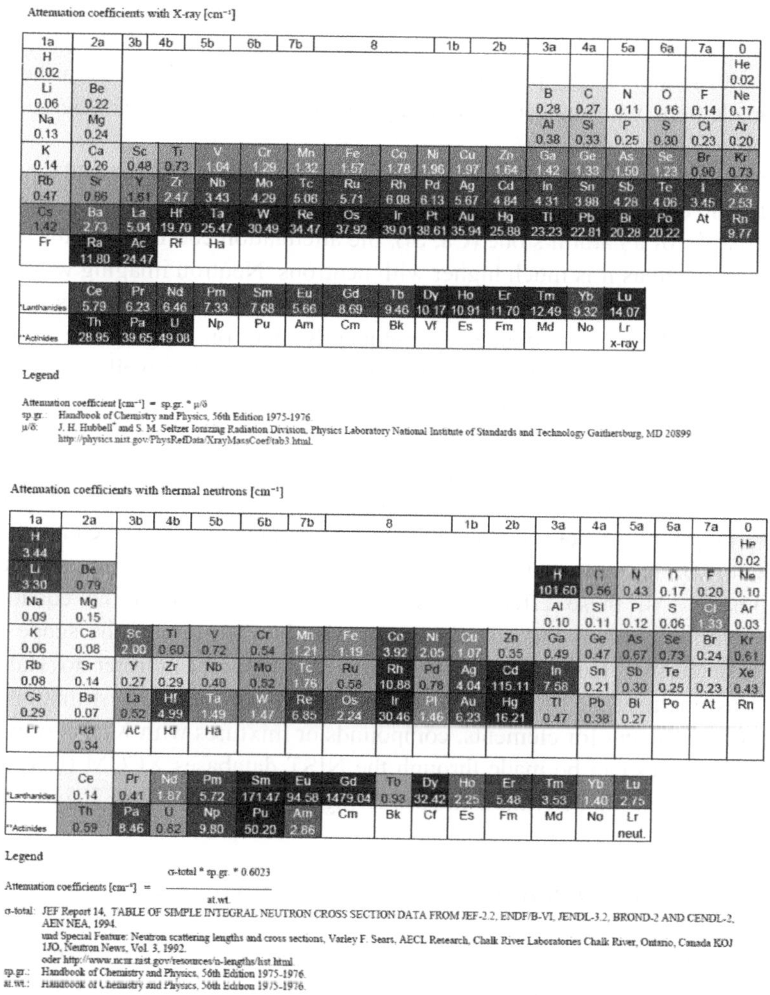

Figure 5.1. *Periodic table with attenuation coefficients of the elements for X-rays at 125 keV and thermal neutrons (Grünauer 2005; De Beer 2018). For a color version of this figure, see www.iste.co.uk/derluyn/salt.zip*

An overview of the attenuation coefficients of the elements is given for X-ray and neutron radiation in Figure 5.1. The attenuation coefficients for X-rays are strongly dependent on the atomic number of the elements. The higher the atomic number, and thus the heavier the element, the higher the interaction probability with an atom of the element. Neutrons have a different interaction behavior. Some light elements like water, boron or chloride have a high interaction probability, whereas some heavy elements like lead have low values, meaning that they are relatively transparent to neutrons. The reason for this lies in the different interaction of X-ray photons and neutrons with matter: X-ray photons interact mostly with the electrons in the atomic shell of an element, whereas neutrons mainly interact with the atomic nucleus.

Therefore, X-ray and neutron imaging are highly complementary. When comparing the attenuation coefficient of hydrogen with the main elements present in the solid matrix of a natural stone (Ca, Si), the attenuation coefficient with X-rays is much lower, whereas it is much higher with neutrons. Neutron imaging will result in images with a larger contrast between water and solid material, whereas water will not be very visible in X-ray images. Consequently, neutron imaging is a well-suited technique to study liquid transport, whereas X-ray imaging is well suited to study the solid phases in a geomaterial.

We remark that equations [5.1] and [5.2] are only valid for pure materials consisting of one element and for monochromatic, that is, mono-energetic, radiation. When working with real porous materials, we penetrate a mixture of elements. Also, most X-ray radiation sources in the laboratory and neutron sources are polychromatic. All these effects need to be accounted for in the equations to correctly employ images for quantitative studies but are out of the scope of this chapter. For further reading on this subject, the reader can refer to Halmshaw (1995), Hassanein et al. (2005) and Derluyn et al. (2019a). An estimation of the attenuation coefficient for elements, compounds or mixtures with X-rays or neutrons at specific energies can be made through the NIST databases XCOM (Berger et al. 2010) and their neutron activation and scattering calculator (Kienzle 2020).

5.2.2. *Imaging of geomaterials and dynamic processes*

The imaging process consists of penetrating the optically opaque material and capturing the transmitted beam intensity on a detector. The detector converts the X-rays or neutrons into visible light, resulting in a 2D-image, a radiograph. The gray-values of the radiograph reflect Lambert–Beer's law:

$$T = \frac{I}{I_0} = e^{-\sum_i \mu_i d_i} \qquad [5.3]$$

where we sum over each component i present in the material, with μ_i the corresponding attenuation coefficient and d_i the corresponding thickness the X-ray path crosses. A radiograph thus represents a transmission image T of the sample. When taking the natural logarithm of this image, we obtain its attenuation image A:

$$A = -\ln T = \sum_i \mu_i d_i \qquad [5.4]$$

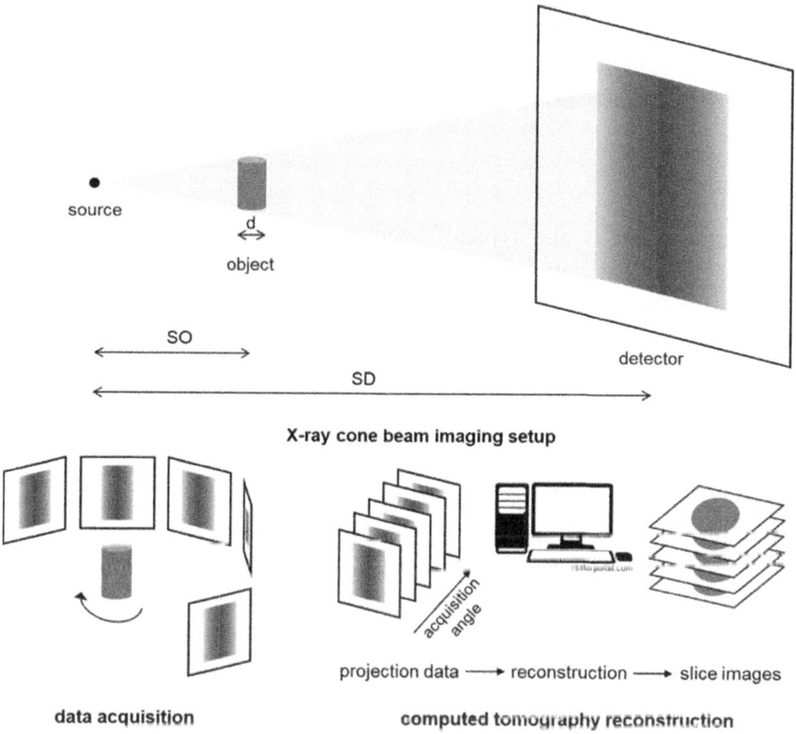

Figure 5.2. *Schematic setup for a cone-beam X-ray imaging setup and the 3D computed tomography reconstruction workflow*

To obtain a 3D reconstruction of the scanned object, multiple radiographs are acquired over different angles. This is typically done by rotating the sample, the source and the detector keeping a fixed position. Using a computational step for the reconstruction, classically based on a backprojection algorithm (Kak and Slaney 1988), tomography slices are calculated from the radiographs, that is, images that are perpendicular to the projection plane (detector plane) as well as to the rotation axis of the object. These tomography slices can then be volume-rendered to obtain a 3D volumetric representation of the object.

A schematic representation of a setup for X-ray imaging using a cone beam setup is given in Figure 5.2, together with a representation of the reconstruction process. In a cone beam setup, the magnification *M* is determined by the geometrical distances between the source and the scanned object (*SO*), and the source and the detector (*SD*): $M = SO/SD$. The pixel size of the 2D radiograph or the voxel size *V* of the 3D reconstruction is then determined by the ratio between the physical pixel size Δx of the detector and the magnification: $V = \Delta x/M$. The number of projection radiographs *P* that need to be taken to obtain a good 3D reconstruction of a sample with a maximal dimension of *d* in the transverse direction, at a voxel size *V*, is defined by the Nyquist theorem (Kharfi 2013):

$$P \geq \frac{\pi}{2}\frac{d}{V} \qquad [5.5]$$

A cone beam setup is the typical setup in laboratory X-ray systems. The geometrical magnification can further be combined with an optical magnification through lenses on the detector side. When using synchroton X-ray radiation or neutron radiation for imaging, the beam is parallel, and the magnification is established purely through lens optics. Synchrotron X-ray imaging and neutron imaging have to be performed at large-scale facilities, such as the European Synchrotron Radiation Facility (ESRF, Grenoble, France), the Synchrotron Soleil (Paris, France) or the Swiss Light Source (SLS, Villigen, Switzerland), and the Institut Laue Langevin (ILL, Grenoble, France) for Neutrons for Society or the Swiss Spallation Neutron Source (SINQ, Villigen, Switzerland). The advantage of X-ray imaging is that it is much more accessible to users as many laboratory facilities exist, whereas neutron imaging can only be performed at large-scale facilities because of the neutron production process through nuclear reactions. A recent development at the neutron imaging beamline D50 at ILL is the combination with a laboratory X-ray system, allowing simultaneous acquisitions of an object with the two imaging modes (Tengattini et al. 2020).

For dynamic studies, the temporal resolution of the imaging system plays an as important role as the spatial resolution. Temporal resolutions down to tens of seconds per tomographic scan can be obtained in specialized laboratory X-ray setups (Bultreys et al. 2016). Higher temporal resolutions require acquisitions at synchrotron sources as they have higher X-ray fluxes. At laboratory X-ray scanners, spatial resolutions in the order of hundreds of nanometers can be obtained, but when very high spatial resolutions are required, synchroton X-ray nanotomography (Villanova et al. 2017) or ptychography (De Boever et al. 2015) needs to be used, allowing to go down to spatial resolutions of tens of nanometer. The spatial resolution of neutron imaging is limited to a few microns (Trtik and Lehmann 2016), whereas researchers at ILL pushed the temporal resolution down to 1.5 s for a neutron tomography at 200 μm spatial resolution, the current record (Tötzke et al. 2019). In general, the following rule holds: when higher temporal resolution is required, one has to sacrifice spatial resolution. This can be seen from equation [5.5], that is, a higher temporal resolution typically requires less projections, which corresponds to a larger pixel size, thus a lower spatial resolution for the same size of sample.

In the next two sections, we will illustrate the potential of dynamic radiography and tomography for the study of sodium chloride crystallization in geomaterials. We will start with a neutron radiography study performed at the ICON beamline at SINQ (Kaestner et al. 2011), and follow-up with an X-ray tomography study performed at the HECTOR scanner of the Ghent University Centre for X-ray Tomography (Masschaele et al. 2013).

5.3. Saline water evaporation and crystallization-induced deformations studied with high-resolution neutron radiography

5.3.1. *Experimental procedure and image analysis*

A sample of Savonnières limestone with a horizontal cross-section of 10×10 mm^2 and 8.25 mm in height is treated with a hydrophobic agent (SILRES® BS 280, Wacker) applied on the top side by painting this side four times with a paint brush. The sample is sealed on the lateral sides by a liquid and vapor tight membrane (aluminum tape) in order to ensure that no moisture transport occurs through these sides, as illustrated in Figure 5.3. The hydrophobic treatment forms a fine network of silicon resin on the interior pore wall, making the pore surface water repellent but not blocking the pores. In this way, capillary action in the upper zone of the samples is suppressed, but drying, by water vapor diffusion, can still occur through the pore space. The hydrophobic treatment is applied to prevent salt efflorescence at the top surface and to induce in-pore crystallization. As will be seen below, the hydrophobic product is not deposited uniformly within this upper zone but the deposition actually leads to a gradient in this top layer.

The sample undergoes three wetting-drying cycles, first with pure water and two subsequent cycles with a 5.8 molal NaCl solution. The sample is wetted by capillary uptake by bringing the bottom side of the samples into contact with water or salt solution, until a stable mass is obtained. The bedding direction of the stone is perpendicular to the one-dimensional (1D) transport. Wetting by capillary uptake is done at room conditions (21°C and 22% RH). After capillary saturation, that is, after about 2.5 h, the sample is sealed on the bottom side by aluminum tape. Consequently, drying can only occur through the hydrophobically treated top side.

Drying takes place at 45°C and 5% RH in a closed test chamber in the neutron imaging beamline. Radiographs are acquired during drying of the sample at 3.5 min intervals, over a period of 10 h. The exposure time is 90 s per radiograph. The pixel size of the acquired images is 13.5 μm. After the neutron image acquisition, the sample is removed from the neutron test setup and further oven-dried at 50°C and 3% RH until the fully dry state is reached before starting a new cycle. A neutron radiograph is taken of the fully dried sample at the end of the second wetting-drying cycle with salt solution (Derluyn 2012; Derluyn et al. 2019b).

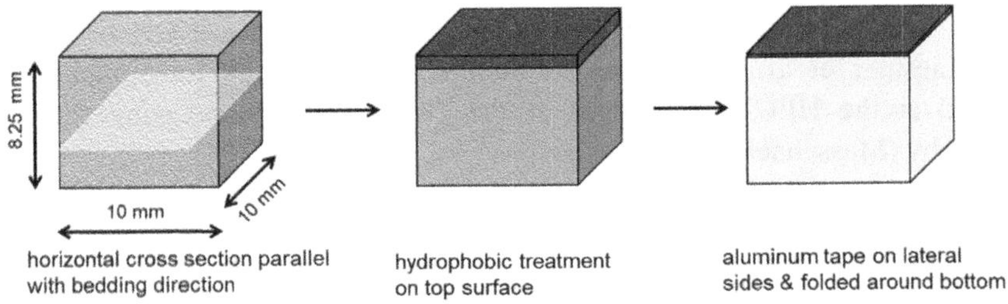

Figure 5.3. *Sample preparation for drying experiments*

By applying Beer–Lambert's law on the neutron radiographs, the moisture content w (kg/m^3) at time t is calculated pixel-wise from the images as follows:

$$w(t) = \frac{\rho_l \cdot d_l(t)}{d_s} = \frac{\rho_l}{\mu_{l,N} \cdot d_s}\left(A_{wet}(t) - A_{dry}\right) \quad [5.6]$$

with ρ_l the density of the liquid phase, that is, water or salt solution, d_l the thickness of the liquid the X-ray path crosses and d_s the total thickness of the sample. $\mu_{l,N}$ is the attenuation coefficient with neutrons for the liquid phase, and A_{wet} the attenuation image of the sample while drying, A_{dry} is the attenuation of a fully dry sample. The attenuation image and the resulting moisture content quantification for the sample after the first saturation with NaCl solution is given in Figure 5.4. As the drying process is considered mainly 1D (in the height direction), 1D moisture profiles are obtained by calculating the average moisture content at every height position, over the central one-third of the width of the sample, as indicated in Figure 5.4 by the white lines.

The total decrease in mass of the sample while drying is consequently expressed as:

$$\Delta m(t) = \sum_{i=1}^{n} \frac{\rho_l \cdot R_{pxl}}{\mu_{l,N}} A_{wet,i}(t) - \sum_{i=1}^{n_0} \frac{\rho_l \cdot R_{pxl}}{\mu_{l,N}} A_{wet,i}(t_0) \quad [5.7]$$

with n_0 the number of pixels corresponding to the area occupied by the sample at time t_0, that is, at the start of drying, and n the number of pixels corresponding to the area occupied at time t. R_{pxl} is the area of one pixel (13.5^2 μm²/pixel).

The global deformation induced during the drying process is quantified in 2D by counting the number of pixels covering the sample for every image during the drying test and multiplying this value by the pixel area. The global deformation is then represented by the change in area R as:

$$\frac{\Delta R}{R} = \frac{R(t) - R(t_0)}{R(t_0)} \qquad [5.8]$$

Neutrons are well attenuated by chlorine, much more than by the elements of the limestone (see also Figure 5.1). Therefore, the precipitated mass of NaCl crystals per unit of volume $m_{salt,v}$ can be quantified from the radiograph taken once the sample is fully dried out, corresponding to the attenuation image $A_{dry+salt}$ (Derluyn et al. 2013):

$$m_{salt,v} = \frac{\rho_{salt} \cdot d_{salt}}{d_s} = \frac{\rho_{salt}}{\mu_{salt,N} \cdot d_s}\left(A_{dry+salt} - A_{dry}\right) \qquad [5.9]$$

with ρ_{salt} the density of the salt crystal and $\mu_{salt,N}$ the attenuation coefficient with neutrons for the salt crystal.

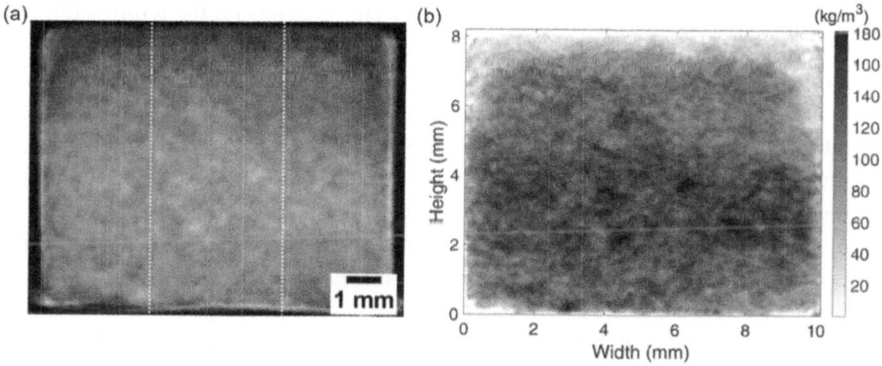

Figure 5.4. *The (a) attenuation image and (b) moisture content quantification for the sample after a first saturation with NaCl solution. The white dotted lines are the boundaries of the central area over which the average moisture content is calculated at every height position to obtain 1D drying profiles as in Figure 5.5*

5.3.2. *Drying-induced salt crystallization in porous limestone*

5.3.2.1. *Drying*

During evaporative drying, the drying rate is initially determined by the difference in vapor pressure between the environment and the moisture front, and by the diffusion resistance. The drying rate is expressed as:

$$q = \beta \cdot \left(p_{v,env} - p_{v,MF}\right) \qquad [5.10]$$

with $p_{v,env}$ the vapor pressure of the environment, which depends on the temperature T_{env} and the relative humidity RH_{env} and is defined as:

$$p_{v,env} = p_{v,sat} \cdot RH_{env} \qquad [5.11]$$

where $p_{v,sat}$ is the saturated vapor pressure, which is a function of the temperature. $p_{v,MF}$ is the vapor pressure at the position of the moisture front, defined as:

$$p_{v,MF} = a_{w,MF} \cdot p_{v,sat} \cdot e^{\frac{p_{c,MF}}{\rho_w R_v T_{MF}}} \qquad [5.12]$$

with $a_{w,MF}$ the water activity at the moisture front, $p_{c,MF}$ the capillary pressure at the moisture front, ρ_w the water density, R_v the gas constant for water vapor and T_{MF} the temperature at the moisture front. The water activity depends on the temperature and the salt concentration and accounts for the change of the vapor pressure in equilibrium with a salt solution, compared to the vapor pressure in equilibrium with pure water. For pure water, the water activity equals 1. The higher the salt concentration gets, the lower the water activity will be, leading to lower vapor pressures. The capillary pressure equals the pressure difference across the liquid–gas interface formed by the menisci in an unsaturated porous medium and is defined as: $p_c = p_l - p_g$, with p_l the liquid pressure and p_g the gas pressure (which is assumed here constant and equal to the atmospheric pressure).

β is the total vapor transfer coefficient, which is generally expressed as:

$$\beta = \frac{1}{\frac{1}{\beta_{env}} + \int_0^{X_{MF}} \frac{\partial x}{\delta_v(x)}} \qquad [5.13]$$

where the denominator represents the total vapor diffusion resistance. β_{env} is the convective vapor transfer coefficient, defining the vapor exchange between the material surface and the environment. This coefficient depends on several parameters such as the air velocity and the thickness of the boundary layer above the drying surface. When the moisture front is at the outer surface, the total vapor transfer coefficient equals β_{env}. When the initial moisture front is located at a certain position below the outer surface, an additional diffusion resistance is present, reducing the drying rate. This resistance is expressed by the second term in the denominator, where δ_v is the vapor permeability of the porous medium located between the moisture front and the outer surface. Such a resistance is present when a material is covered by a hydrophobic material (e.g. hydrophobic plaster on brick) or when the upper volume of a material is hydrophobically treated (Poupeleer 2007). δ_v can be the location dependent on the degree of hydrophobicity. Also, the presence of salt crystals blocking the pores can cause a reduction in δ_v.

As long as the moisture transport from inside the material is in equilibrium with the vapor transport from the front to the exterior, the moisture front will remain at the same position. This is defined as the first drying stage, and the drying rate is determined by equation [5.10]. For pure water drying, this period is characterized as the constant rate period, the vapor pressure reduction due to capillary pressure changes is insignificant during this period (Scherer 1990). For salt solution drying, the water activity will affect the drying rate (equation [5.12]). Once the flow toward the moisture front position is no longer maintained, the moisture front will recede deeper into the material, and the drying rate slows down, evolving into the second stage of drying. In this stage, the material properties, such as the moisture diffusivity, determine the further evolution of the moisture distribution in the material (Pel et al. 2002).

Figures 5.5(a) and (b) depicts the moisture profiles while drying after saturation with pure water and after the first saturation with a 5.8 molal NaCl solution. The penetration depth of the hydrophobic treatment is estimated from the initial moisture profile, which corresponds to the capillary saturated state. The height of the hydrophobic zone is indicated by the arrow and is estimated from the position at which the initial moisture content shows a continuous decrease with increasing sample height. The hydrophobic treatment penetrated the upper part of the stone over a depth of about 3 mm. The initial moisture profile shows that the hydrophobic treatment did not make the upper volume of the stone fully hydrophobic. It creates a gradual reduction of the wettability of the pore space in the penetration zone, that is, from no capillary action close to the top side to full capillary action at the penetration depth of approximately 3 mm.

The drying experiment with water allows us to characterize the duration of the first drying period and the onset of the second drying period. The moisture profiles in Figure 5.5(a) show that there is initially a quite constant decrease of the moisture content over the height of the sample, and the moisture front remains at the same position. The drying rate is constant, as can be deduced from the linear character of the mass decrease during the first 360 min (blue dashed line in Figure 5.5(c)).

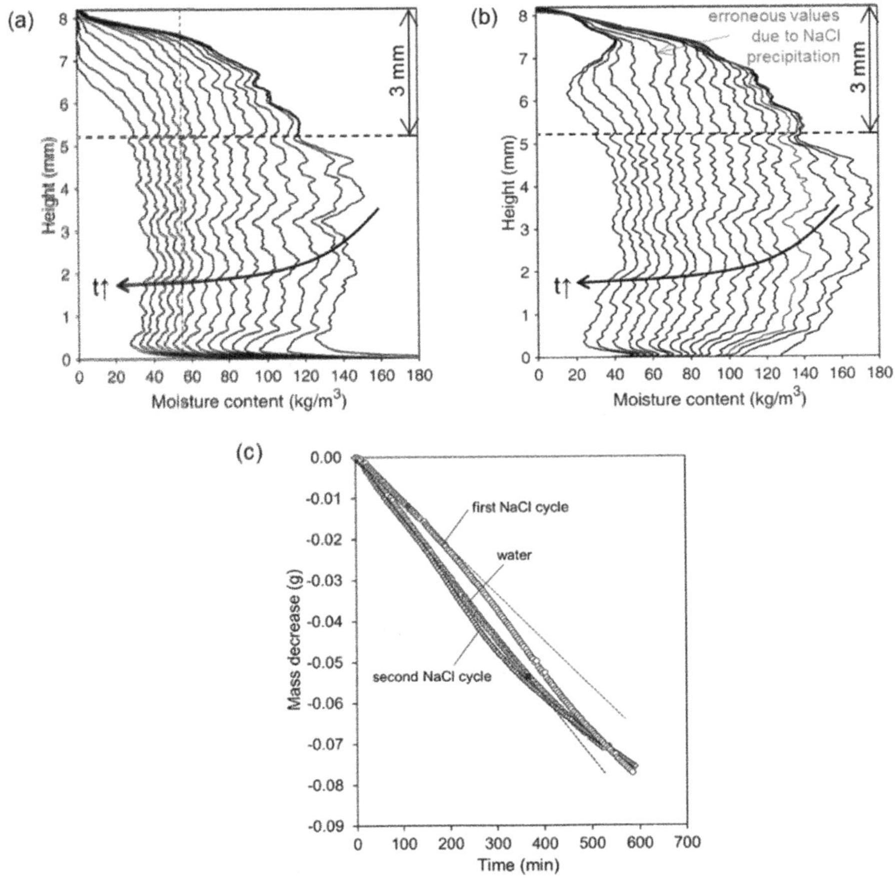

Figure 5.5. *Drying profiles at time intervals of 37 min for the sample saturated with (a) water (Reproduced by permission from Springer Nature Customer Service Centre GmbH: Springer, Transp. Porous Media, 128, 895–913, Copyright (2019)), and (b) saturated with 5.8 molal NaCl solution. The hydrophobic zone is indicated by the arrow. (c) Mass decrease during drying of sample saturated with water and during drying of first and second wetting-drying cycle with 5.8 molal NaCl solution. For a color version of this figure, see www.iste.co.uk/derluyn/salt.zip*

Due to the hydrophobic treatment, the moisture content is lower in the treated zone than the capillary moisture content of the non-treated limestone below the hydrophobic zone. However, it can be seen from the moisture profiles that a hydraulic connection is maintained almost up to the outer surface during the first stage of drying. This hydraulic connection is heterogeneously distributed, as we likely have only localized capillary rise up to the surface. It is only when the moisture content in the lower part of the sample (non-hydrophobically treated stone) has decreased to around 55 kg/m^3 (indicated in blue in Figure 5.5(a)) that the hydraulic connectivity is no longer guaranteed and that the moisture front recedes into the sample, characterizing the second stage of drying. The drying rate at this point slows down, as the total vapor transfer coefficient decreases (equation [5.13], blue point in Figure 5.5(c)).

Comparing Figure 5.5(b) with Figure 5.5(a), it can be concluded that the drying rate of the sample capillary saturated with a 5.8 molal NaCl solution is similar to the drying rate of the sample capillary saturated with water. This is confirmed by the results of the total mass decrease (Figure 5.5(c)), and even after a second wetting-drying cycle, the drying rate remains quasi the same. However, we would expect a decrease in drying rate between a water-saturated sample and a NaCl-solution-saturated sample, as the water activity of the NaCl solution is lower (0.77 for a 5.8 molal concentration at 45°C). The linear slope of the mass decrease during the first 100 minutes of drying reduces with 24% for the first wetting-drying cycle, similar to this water activity decrease (red dashed line in Figure 5.5(c) up to the red point). The mass decrease then however accelerates, which can be explained by the creation of additional transport paths through crack-formation, increasing the drying rate. This is confirmed by the post-mortem analysis of the sample by an X-ray tomography scan, as shown in Figure 5.7(b). The sample thus cracks, which is due to the accumulation of NaCl and the associated force of crystallization. In addition, the subflorescence that has formed is in itself a porous, strongly hydrophilic, porous medium, which might form additional transport paths as well, enhancing the drying (see also Chapter 3).

5.3.2.2. *Salt accumulation*

1D transport of salt species in the liquid phase in a porous medium with a constant total porosity is described by the mass balance:

$$\frac{\partial(\rho_l S_l C)}{\partial t} + \frac{\partial(S_{cr}/V_{m,cr})}{\partial t} = \frac{\partial}{\partial x}\left(\rho_l S_l \frac{D}{\tau}\frac{\partial C}{\partial x}\right) - \frac{\partial}{\partial x}(\rho_l S_l v_l C) \qquad [5.14]$$

with ρ_l the liquid density, S_l the volumetric liquid saturation degree, C the salt concentration, S_{cr} the volumetric crystal saturation degree and $V_{m,cr}$ the molar volume of the salt crystal. D is the diffusion coefficient for the salt species in the liquid phase and τ is the tortuosity of the porous medium. v_l is the velocity of the liquid phase. The first term on the right-hand side represents the diffusive ion flow, the second term represents the advective ion flow. As long as no crystallization occurs, the second term on the left-hand side equals zero. The change in salt concentration is determined by a competition between the advective and diffusive flow. When the advective flow is dominant, the concentration will increase in the zone where the liquid flows to, and the concentration distribution will be heterogeneous throughout the sample. When the diffusive flow is dominant, salt will be transported from zones of high concentration toward zones of lower concentration and the concentration will equilibrate, leading to a homogeneous distribution of the salt concentration. This competition between the advection and diffusion can be characterized by the dimensionless Péclet number (Huinink et al. 2002):

$$Pe = \frac{|v_l|L}{D/\tau} \qquad [5.15]$$

where L is the characteristic length scale. When $Pe >> 1$, the salt transport is advection dominated; for $Pe << 1$, the transport is dominated by diffusion.

The liquid velocity can be estimated from the experimentally determined moisture profiles, starting from the following mass balance expression for the temporal evolution of the moisture content w:

$$\frac{\partial w}{\partial t} = -\frac{\partial}{\partial x}(w \cdot v_l) \qquad [5.16]$$

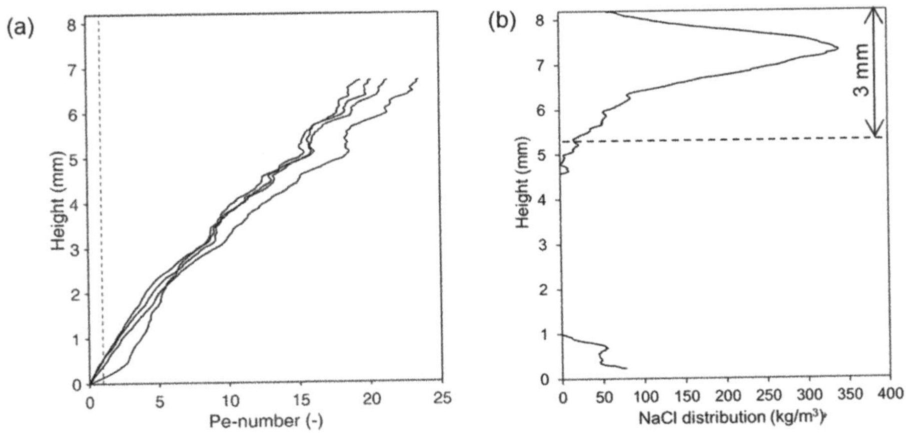

Figure 5.6. *(a) Peclet number over the height of the sample at time intervals of 37 min during the first 150 min of drying, after the first uptake of 5.8 molal NaCl solution (derived from Figure 5.5(b)). The vertical dashed line represents a Peclet number of 1 (Reproduced by permission from Springer Nature Customer Service Centre GmbH: Springer, Transp. Porous Media, 128, 895–913, Copyright (2019)). (b) Halite profile determined from the neutron radiograph of the sample dried out after two wetting-drying cycles. The hydrophobic zone is indicated by the arrow*

The velocity is found by integrating both sides over the spatial variable x and assuming that at $x=L$ the velocity is zero, as the sample is sealed on that side (Petkovic 2005):

$$v_l(x) = \frac{1}{w(x)}\frac{\partial}{\partial t}\int_x^L w(x')dx' \qquad [5.17]$$

Precipitation is expected to be rapid during the first cycle for the NaCl-solution saturated sample due to its high initial concentration, which amounts to 93% of the saturated concentration at 45°C. Figure 5.6(a) shows the Peclet number during the first 150 min of drying, that is, before the drying rate increase, using a value of 1.58×10^{-9} m²/s for the diffusion coefficient (Rard and Miller 1979) and 24 for the tortuosity (Derluyn 2012). The Peclet number is clearly higher than 1 (the dashed vertical line in the figure corresponds to Pe = 1), meaning that the ion transport is governed by advection and that salt will accumulate at the moisture front. As the Savonnières sample shows a long first drying stage, most of the salt is thus expected to precipitate in the upper volume. This is confirmed by the salt distribution quantified from the neutron radiograph at the end of the second wetting-drying cycle, shown in Figure 5.6(b). Salt precipitation takes place in the upper, partly-hydrophobic volume, showing a local peak of high accumulation.

This precipitation is also reflected in the profiles of Figure 5.5(b). As the neutron imaging does not discriminate between sodium chloride in solution and sodium chloride in its crystal phase, the attenuation due to precipitated salt is attributed to salt solution during the moisture content quantification. This leads to an overestimation of the moisture content in the hydrophobic zone as the sample dries out.

5.3.2.3. *Deformation*

Under isothermal conditions and for a freely deformable medium, deformations will be induced by two counteracting phenomena, namely drying shrinkage and crystallization-induced dilation. Shrinkage is related to a decreasing liquid saturation degree of the pore space coupled with a decreasing capillary pressure, and dilation is related to an increasing pore filling by salt crystals coupled with the crystallization pressure exerted by them. Under the assumption of small displacements, the hygric and crystallization strains lead to a volumetric strain:

$$\frac{\Delta V}{V} = \varepsilon_{hygr} + \varepsilon_{cryst} \qquad [5.18]$$

The distribution of liquid and salt crystals throughout the sample will determine the homogeneity of the volumetric strain field over the global volume of the sample, and the possible development of strain localization. The global change in the sample's volume then results from the accumulation of the local volumetric strains.

In rock material, strain localization is the prelude for material damage, at which the assumption of small displacements does not longer hold. The creation of crack planes is associated with local jumps in the displacements, which result in a strong increase of the global volumetric strain.

The radiographs only allow estimating the global strain in two dimensions, following equation [5.8]. The result is given in Figure 5.7(a) for the drying phase of the first and second wetting-drying cycle with 5.8 molal NaCl solution. During the first cycle, the deformation initially increases slowly, to values in the order of 1‰, implying small dilations, and thus the salt crystallization compensates the hygric shrinkage. After more than 100 min however, the deformation starts to increase significantly, reaching a global deformation of over 1%. The sharp increase in deformation corresponds to the displacement jumps induced by cracking, and the timing is congruent with the changing slope of the mass decrease curve in Figure 5.5(c).

During the second cycle, large deformations occur from the start of the curve. Halite crystals, precipitated during the prior cycle, are present and facilitate the nucleation of new crystals, leading to the immediate continuation of the crystallization-induced deformation as soon as drying starts. Cracks that are already present are expected to open further. In addition, new cracks can be created, leading to the high deformation values reported in Figure 5.7(a).

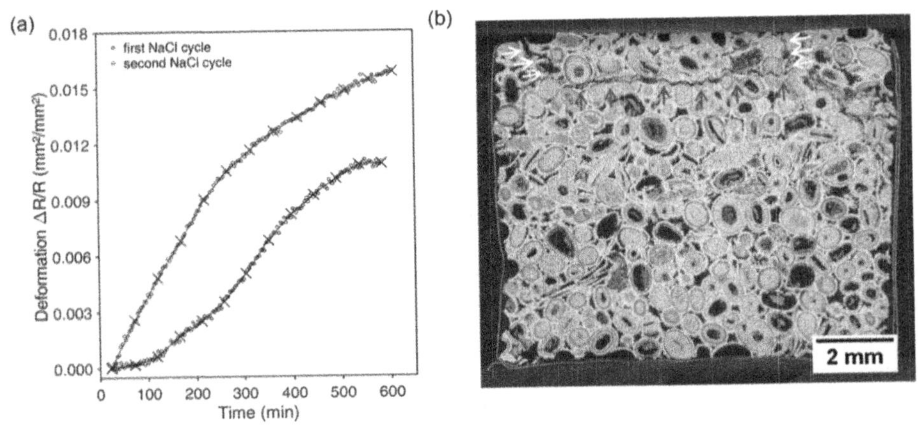

Figure 5.7. *(a) Global deformation $\Delta R/R$ during the drying phase of the first and second wetting-drying cycle after capillary saturation with a 5.8 molal NaCl solution. (b) Vertical cross-section of a reconstructed volume of a sample obtained through X-ray tomographic microscopy after NaCl precipitation. The blue arrows indicate the main horizontal crack plane, coinciding with the halite precipitation zone. The white arrows indicate additional cracks enhancing vapor transport to the exterior. For a color version of this figure, see www.iste.co.uk/derluyn/salt.zip*

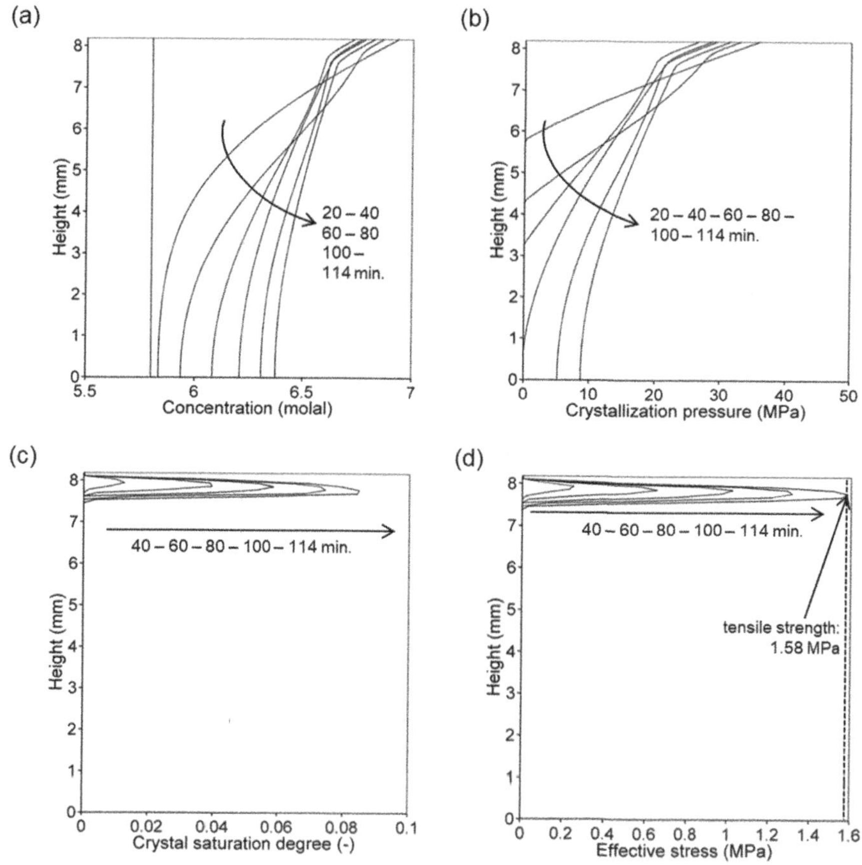

Figure 5.8. *(a) Concentration, (b) crystallization pressure, (c) crystal saturation degree and (d) effective stress over the height of the sample (reprinted from J. Mech. Phys. Solids, 63, 242–255, Copyright (2014), with permission from Elsevier))*

Prior to crack formation, we can use the poroelastic theory, as developed in the previous chapter, to describe the deformation of the sample. To that extent, a coupled thermo-hydro-chemo-mechanical numerical model was developed (Derluyn et al. 2014). The model confirmed that for the sample described here, that is, a freestanding porous material that is subjected to environmental changes, the effective stress σ in the material can be approximated by accounting only for the crystallization-induced stress:

$$\sigma = bS_{cr}p_{cr} \quad [5.19]$$

where b is the Biot coefficient, S_{cr} is the crystal saturation degree (volume fraction of the pore volume filled with crystals) and p_{cr} is the crystallization pressure exerted on the pore walls, as expressed in Chapter 2 by equation [2.4]. The numerical model allows tracing all the physical variables inside the sample, such as the concentration, the crystallization pressure, the amount of salt crystals and the effective stress, depicted in Figure 5.8.

Due to the drying process, the concentration increases, and due to the high Peclet number, higher concentrations are obtained in the upper volume of the sample. As such, supersaturation is being generated, which corresponds to a certain crystallization pressure. However, this pressure will only be exerted on the pore walls if crystals effectively precipitate. The model assumes that NaCl precipitates when the salt concentration exceeds 10% of the concentration at saturation for NaCl (Flatt 2002). Crystals thus form in the upper volume of the sample, in line with the experimental observations, and the effective stress is only being build up in this zone.

Cracking is initiated when a certain damage criterion is exceeded. Different damage criteria exist, in the example developed here (Figure 5.8(d)), cracking occurs when the effective stress exceeds the tensile strength σ_T measured by a uniaxial tensile strength test:

$$\sigma \geq \sigma_T \qquad [5.20]$$

Other authors use a strain energy criterion, predicting damage when the strain energy imposed by the salt crystals exceeds the failure strain energy found in a uniaxial tensile strength test (Espinosa-Marzal et al. 2011), resulting in the following constraint:

$$\sigma \geq \frac{\sigma_T}{\sqrt{3(1-2\nu)}} \qquad [5.21]$$

with ν the Poisson coefficient.

The simulation results presented in Figure 5.8 depend on the parameters chosen for the crystal growth. The parametric study that has been performed can be found in detail in Derluyn et al. (2014). The results presented here are for the parameters that result in damage after 114 min, congruent with the timing at which the global deformation increases significantly. The predictive power of thermo-hydro-chemo-mechanical simulations for salt damage risk assessment thus strongly depends on the accuracy of the incorporated salt growth models. The kinetic parameters for salt growth inside porous media are however difficult to access, and future research should therefore focus on experimental and/or modeling studies for the reliable prediction of these parameters.

5.4. Salt crystallization dynamics in sandstone studied with 4D laboratory X-ray micro-tomography

5.4.1. *Experimental procedure and image analysis*

As evoked in Chapter 3, during drying-induced crystallization, the distribution of salt crystals in the pore space (subflorescence) and on the material's surface (efflorescence) strongly influences the drying rate. Desarnaud et al. (2015) described the effect of the crystal growth on the drying behavior of Mšené sandstone initially saturated with 6.1 molal NaCl solution (i.e. a saturated NaCl solution), by comparing drying at 20% RH and 50% RH, at room temperature. These RH values are representative for winter and summer conditions for the salt NaCl, which is not temperature sensitive. By means of balance measurements, primary drying at 20% RH was found to take longer than at 50% RH. This is counterintuitive, as we would expect a larger drying rate at lower RH (equation [5.10]). The slower drying is attributed to the formation of a salt skin on the surface at 20% RH (crusty efflorescence).

To elucidate on the coupled drying-crystallization dynamics in the porous sandstone, drying and crystal growth are visualized and quantified simultaneously using time-resolved X-ray micro-tomography (4D µCT) (Derluyn et al. 2016). Two cylindrical samples of 8 mm in diameter and 10 mm in height are cored from a Mšené sandstone plate. The samples are scanned in their dry state, prior to capillary saturating them by immersion in a saturated NaCl solution for 30 min. The samples are then scanned in their wet state, and subsequently during their drying, by scanning them every 30 min during the first 3 h, and every hour during the succeeding 12 h. Drying is controlled by placing the sample in a custom-built climatic chamber, compatible with the µCT setup. Conditioned air is blown in the test chamber at a slow rate (3.3 mL/s), ensuring that the samples are dried at a constant temperature of 19.2°C and a RH of 20% and 50% for the first and the second sample, respectively.

For each scan, a total of 1351 projections are acquired over an angle of 360°, with an exposure time of 400 ms per projection. The X-ray tube provides a voltage of 100 kV with a target power of 10 W. The source-detector distance is 1,530.7 mm and the source-object distance 38.3 mm. The projections are binned by a factor 2 × 2 in order to increase the signal-to-noise ratio, resulting in a reconstructed voxel size of 10 µm. The raw data are reconstructed using the software Octopus Reconstruction (Inside Matters bvba; Vlassenbroeck et al. 2007). Small shifts in the reconstructed volumes with respect to the reference volume are corrected by aligning all volumes to the reference scan using the software DataViewer (Bruker microCT). This ensures the same positioning in 3D space for all datasets.

Each scan results in a 3D volume of the upper 6 mm of the sample. Due to the spatial resolution aimed for, the complete height of the cylindrical samples could not be imaged in the field of view defined by the HECTOR setup. The 3D volumes are further analyzed with the software AVIZO (FEI). By working with differential images (Boone et al. 2014), the salt solution in the sample and the crystal precipitation are segmented, that is, the voxels belonging to the same phase are identified, allowing to simultaneously obtain quantitative data on the drying and the precipitation kinetics. It needs to be remarked that the quantification based on the 4D μCT imaging is limited in accuracy by the spatial resolution of 10 μm and by the image analysis procedure, and as such can never be as precise as raw balance measurements to assess the drying kinetics. The added advantage however is that these dynamic imaging datasets also provide data on the crystallization kinetics, whereas balance measurements only quantify the drying kinetics.

5.4.2. *Drying-crystallization kinetics*

Representative visualizations of the sample drying at 20% RH are given in Figure 5.9. The drying quantification of the two samples is given in Figure 5.10(a). We observe that drying at 20% RH is initially faster than at 50% RH, due to the higher difference in RH between the sample's surface and the environment. This first drying stage is characterized by a constant rate, due to the capillary action, which induces a hydraulic connection between the surface and the inner pore space. Subsequently, the evaporation at the surface mainly results in salt efflorescence. At 50% RH, this constant rate period lasts until the sample is almost completely dried out. At 20% RH, the drying changes to a slower, nonlinear regime after about 1 h, that is, after approximately half of the initial salt solution has dried out. The drying kinetics found here corroborate the drying kinetics reported by Desarnaud et al. (2015).

The differences in drying kinetics are caused by the differences in salt precipitation dynamics, which are presented in Figures 5.10(b) and (c). The crystal precipitation develops slower at 50% RH, where the salt crystals forming on the surface (efflorescence) and the salt crystals precipitating in the pores close to the surface develop at the same relative rate during the whole duration of drying (with respect to their respective total amounts, Figure 5.10(c)). No salt crystals are found deep inside the volume. While drying at 20% RH, the crystal growth at the surface develops first and much faster than at 50% RH, mainly during the first hour of the experiment, and then reaches a plateau. The salt crystals precipitating just below the surface at this lower RH develop more gradually, during the whole duration of the drying. The period of the formation of efflorescence at 20% RH is congruent with the period of the constant drying rate in the drying curve. The subsequent slower drying stage at 20% RH must thus be resulting from this salt efflorescence, corroborating the statement that a salt skin is forming on the surface during the first stage of drying (Desarnaud et al. 2015), partially closing the pores and causing a slower drying during the continuation of the process. The salt skin still grows slightly during this second period, but most of the salt crystals precipitate just below this layer, in the subsurface.

Figure 5.9. *Snapshots of 3D visualizations of the Mšené sandstone sample while drying at 20% RH based on X-ray micro-tomography reconstructions. For a color version of this figure, see www.iste.co.uk/derluyn/salt.zip*

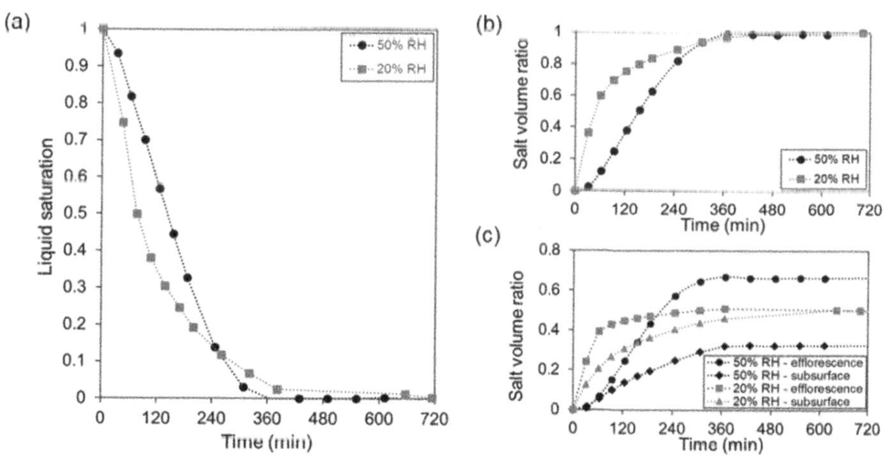

Figure 5.10. *The (a) drying and (b and c) crystallization dynamics of Mšené sandstone at 20% RH and 50% RH. Figure (c) distinguishes between the salt crystals forming on the outer surface (efflorescence) and the salt crystals precipitating just below in the pores of the subsurface*

Grementieri et al. (2017) used this experimental dataset to develop a numerical model including salt crust formation. In brief, they developed a coupled hygro-chemical model, in the same spirit as the one by Derluyn et al. (2014), and parameterized the vapor transfer coefficient in the drying rate expression (equation [5.10]) to include the effect of the salt skin. Note that this is only a phenomenological, and thus not an explicit numerical description, of the salt skin. The experimental and simulation data for the drying and crystallization kinetics are represented in Figure 5.11. When drying at 50% RH, the same vapor transfer coefficient as for drying a sample saturated with pure water can be employed. Thus, the numerical simulation confirms that at 50% RH the salt precipitation, resulting in patchy efflorescence, does not affect the drying rate. At 20% RH, the vapor transfer coefficient decreases as a function of the amount of salt precipitation, reflecting the formation of the salt skin in a numerical way.

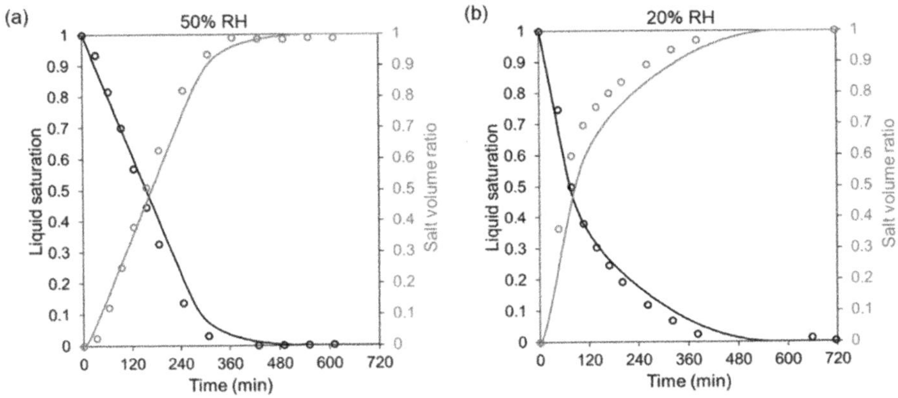

Figure 5.11. *Experimental results from quantitative X-ray image analysis (circle symbols) and numerical results (solid lines) for the drying and crystallization kinetics of Mšené sandstone when wetted with saturated NaCl solution and dried at (a) 50% RH and (b) 20% RH*

5.5. Conclusion

X-ray and neutron imaging provide us with a look inside geomaterials in a non-destructive way. By employing image analysis techniques, quantitative data are obtained on salt transport and precipitation dynamics, as well as on the mechanical effects of these processes on geomaterials. These techniques prove to be very powerful, and continuous developments pushing the limits in terms of spatial and temporal resolution at X-ray and neutron imaging facilities promise further advancements in studying dynamic processes, like salt crystallization in porous media, in the near future.

The ultimate goal of these types of experiments is to fully comprehend the complex problem of salt precipitation and salt damage in geomaterials in order to predict salt damage risks and to aid in the design of preventive or conservation measures. Understanding the coupling between transport, precipitation and mechanics is defined as the key to assess salt damage risks in porous media such as natural building stones and cultural heritage, and thus to assess their durability. The durability of a stone consists of two stages (Flatt et al. 2017). In the first stage, the accumulation stage, a sufficient amount of salt deposit is being built up. After reaching a certain threshold, damage is initiated, and the second stage sets in, where we get a propagation of damage in the material.

The example discussed in section 5.3 covers these two stages, and by combining the experimental results with a thermo-hydro-chemo-mechanical numerical model, the full history of the sample can be reconstructed. The numerical model thus aids in interpreting the complex coupled phenomena that are ongoing, and to retrieve physical properties such as the crystallization pressure that is being built up. The model output depends of course on the input that it is given, and we pointed out that more data on salt growth kinetics inside porous media would be helpful in improving the predictive potential of such models. However, when used rigorously and within the limits of their validity, these numerical models can be employed with confidence to predict salt damage risks or to simulate the effect of potential remediation strategies.

The second example discussed in section 5.4 focuses on the first stage, the accumulation stage. By means of 4D X-ray micro-tomography, the dynamics of drying and salt crystallization are retrieved simultaneously. This is a first step in how time-resolved X-ray imaging can bridge the gap on getting more data on salt growth kinetics in porous media. The spatial detail on the salt efflorescence is of course not as detailed as what is described in Chapter 3 and illustrates the necessity of combining different experimental techniques to comprehend the beauty and complexity of salt growth in porous media, and the need for joining macroscopic and microscopic views to comprehend this research topic completely.

5.6. Acknowledgments

The research presented in this chapter compiles work performed by the author at different research institutes in collaboration with the research groups of Prof. J. Carmeliet and Prof. V. Cnudde. The author gratefully acknowledges the former financial support by the Department of Architecture at ETH Zurich and by the Research Foundation Flanders (FWO grant 1276814N), and the current financial support from the European Research Council (ERC) under the European Union's Horizon 2020 research and innovation program (grant agreement No. 850853).

For the work presented in section 5.3, special thanks go to Stephan Carl, Roger Vonbank and Jan Hovind for technical support, and Michele Griffa and Peter Moonen for scientific discussions. The work presented in section 5.4 was part of the European JPI-CH project KISADAMA, and warm thanks go to the whole KISADAMA group at Ghent University, University of Amsterdam and University of Bologna for the many fruitful scientific exchanges.

5.7. References

Berger, M.J., Hubbell, J.H., Seltzer, S.M., Chang, J., Coursey, J.S., Sukumar, R., Zucker, D.S., Olsen, K. (2010). XCOM: Photon cross sections database [Online]. Available at: https://www.nist.gov/pml/xcom-photon-cross-sections-database [Accessed 23 April 2020].

Boone, M.A., Nielsen, P., De Kock, T., Boone, M.N., Quaghebeur, M., Cnudde, V. (2014). Monitoring of stainless-steel slag carbonation using X-ray computed microtomography. *Environ. Sci. Technol.*, 48, 674–680.

Bultreys, T., Boone, M.A., Boone, M.N., De Schryver, T., Masschaele, B., Van Hoorebeke, L., Cnudde, V. (2016). Fast laboratory-based micro-computed tomography for pore-scale research: Illustrative experiments and perspectives on the future. *Adv. Water Resour.*, 95, 341–351.

Cnudde, V. and Boone, M.N. (2013). High-resolution X-ray computed tomography in geosciences: A review of the current technology and applications. *Earth-Science Rev.*, 123, 1–17.

Cnudde, V., De Kock, T., Boone, M., De Boever, W., Bultreys, T., Van Stappen, J., Vandevoorde, D., Dewanckele, J., Derluyn, H., Cardenes, V. et al. (2015). Conservation studies of cultural heritage: X-ray imaging of dynamic processes in building materials. *Eur. J. Mineral.*, 27(3), 269–278.

De Beer, F.C. (2018). Neutron and X-ray tomography as research tools for applied research in South Africa. PhD Thesis, North-West University, South Africa.

De Boever, W., Diaz, A., Derluyn, H., De Kock, T., Van Stappen, J., Dewanckele, J., Bultreys, T., Boone, M., De Schryver, T., Skjønsfjell, E.T.B. et al. (2015). Characterization of composition and structure of clay minerals in sandstone with ptychographic X-ray nanotomography. *Appl. Clay Sci.*, 118, 258–264.

Derluyn, H. (2012). Salt transport and crystallization in porous limestone: Neutron-X-ray imaging and poromechanical modeling. PhD Thesis, ETH Zurich.

Derluyn, H., Griffa, M., Mannes, D., Jerjen, I., Dewanckele, J., Vontobel, P., Sheppard, A., Derome, D., Cnudde, V., Lehmann, E. et al. (2013). Characterizing saline uptake and salt distributions in porous limestone with neutron radiography and X-ray micro-tomography. *J. Build. Phys.*, 36, 353–374.

Derluyn, H., Moonen, P., Carmeliet, J. (2014). Deformation and damage due to drying-induced salt crystallization in porous limestone. *J. Mech. Phys. Solids*, 63, 242–255.

Derluyn, H., Boone, M.A., Desarnaud, J., Grementieri, L., Molari, L., de Miranda, S., Shahidzadeh, N., Cnudde, V. (2016). Quantifying salt crystallization dynamics in sandstone using 4D laboratory X-ray micro-CT. In *Science and Art: A Future for Stone: Proceedings of the 13th International Congress on the Deterioration and Conservation of Stone, Volume 1*, Hughes, J.J., Howind, T. (eds). University of the West of Scotland, Paisley.

Derluyn, H., Vontobel, P., Mannes, D., Derome, D., Lehmann, E., Carmeliet, J. (2019a). Saline water evaporation and crystallization-induced deformations in building stone: Insights from high-resolution neutron radiography supplementary information. *Transp. Porous Media*, 128. doi: 10.1007/s11242-018-1151-x.

Derluyn, H., Vontobel, P., Mannes, D., Derome, D., Lehmann, E., Carmeliet, J. (2019b). Saline water evaporation and crystallization-induced deformations in building stone: Insights from high-resolution neutron radiography. *Transp. Porous Media*, 128, 895–913.

Desarnaud, J., Derluyn, H., Molari, L., de Miranda, S., Cnudde, V., Shahidzadeh, N. (2015). Drying of salt contaminated porous media: Effect of primary and secondary nucleation. *J. Appl. Phys.*, 118, 114901.

Espinosa-Marzal, R.M., Hamilton, A., McNall, M., Whitaker, K., Scherer, G.W. (2011). The chemomechanics of crystallization during rewetting of limestone impregnated with sodium sulfate. *J. Mater. Res.*, 26, 1472–1481.

Flatt, R.J. (2002). Salt damage in porous materials: How high supersaturations are generated. *J. Cryst. Growth*, 242, 435–454.

Flatt, R.J., Mohamed, N.A., Caruso, F., Derluyn, H., Desarnaud, J., Lubelli, B., Espinosa-Marzal, R.M., Pel, L., Rodriguez-Navarro, C., Scherer, G.W. et al. (2017). Predicting salt damage in practice: A theoretical insight into laboratory tests. *RILEM Tech. Lett.*, 2, 108–118.

Grementieri, L., Molari, L., Derluyn, H., Desarnaud, J., Cnudde, V., Shahidzadeh, N., De Miranda, S. (2017). Numerical simulation of salt transport and crystallization in drying Prague sandstone using an experimentally consistent multiphase model. *Build. Environ.*, 123, 289–298.

Grünauer, F. (2005). Design, optimization, and implementation of the new neutron radiography facility at FRM-II. PhD Thesis, TU München.

Halmshaw, R. (1995). *Industrial Radiology – Theory and Practice*, 2nd edition. Springer-Science+Business Media, B.V. Chapman & Hall, London.

Hassanein, R., Lehmann, E., Vontobel, P. (2005). Methods of scattering corrections for quantitative neutron radiography. *Nucl. Instruments Methods Phys. Res. Sect. A Accel. Spectrometers, Detect. Assoc. Equip.*, 542, 353–360.

Huinink, H.P., Pel, L., Michel, M.A.J. (2002). How ions distribute in a drying porous medium: A simple model. *Physics of Fluids*, 14, 1389.

Kaestner, A.P., Hartmann, S., Kühne, G., Frei, G., Grünzweig, C., Josic, L., Schmid, F., Lehmann, E.H. (2011). The ICON beamline – A facility for cold neutron imaging at SINQ. *Nucl. Instruments Methods Phys. Res. Sect. A Accel. Spectrometers, Detect. Assoc. Equip.*, 659, 387–393.

Kak, A.C. and Slaney, M. (1988). *Principles of Computerized Tomographic Imaging*. IEEE Press, New York.

Kharfi, F. (2013). Mathematics and physics of computed tomography (CT): Demonstrations and practical examples. In *Imaging and Radioanalytical Techniques in Interdisciplinary Research – Fundamentals and Cutting Edge Applications*, Kharfi, F. (ed.). IntechOpen, London.

Kienzle, P. (2020). Neutron activation and scattering calculator [Online]. Available at: https://www.ncnr.nist.gov/resources/activation/ [Accessed 23 April 2020].

Maire, E. and Withers, P.J. (2014). Quantitative X-ray tomography. *Int. Mater. Rev.*, 59, 1–43.

Masschaele, B., Dierick, M., Van Loo, D., Boone, M.N., Brabant, L., Pauwels, E., Cnudde, V., Van Hoorebeke, L. (2013). HECTOR: A 240kV micro-CT setup optimized for research. *J. Phys. Conf. Ser.*, 463, 012012.

Pel, L., Landman, K.A., Kaasschieter, E.F. (2002). Analytic solution for the non-linear drying problem. *Int. J. Heat Mass Transf.*, 45, 3173–3180.

Pel, L., Pishkari, R., Casti, M. (2018). A simplified model for the combined wicking and evaporation of a NaCl solution in limestone. *Mater Struct.*, 51, 66.

Perfect, E., Cheng, C.L., Kang, M., Bilheux, H.Z., Lamanna, J.M., Gragg, M.J., Wright, D.M. (2014). Neutron imaging of hydrogen-rich fluids in geomaterials and engineered porous media: A review. *Earth-Science Rev.*, 129, 120–135.

Petkovic, J. (2005). Moisture and ion transport in layered porous building materials: A nuclear magnetic resonance study. PhD Thesis, T.U. Eindhoven.

Poupeleer, A.S. (2007). Transport and crystallization of dissolved salts in cracked porous building materials. PhD Thesis, K.U. Leuven.

Rard, J.A. and Miller, D.G. (1979). The mutual diffusion coefficients of $NaCl\text{-}H_2O$ and $CaCl_2\text{-}H_2O$ at 25 C from Rayleigh interferometry. *J. Solution Chem.*, 8, 701–716.

Scherer, G.W. (1990). Theory of drying. *J. Am. Ceram. Soc.*, 73, 3–14.

Tengattini, A., Lenoir, N., Andò, E., Giroud, B., Atkins, D., Beaucour, J., Viggiani, G. (2020). NeXT-Grenoble, the neutron and X-ray tomograph in Grenoble. *Nucl. Inst. Methods Phys. Res. A*, 968, 163939.

Tötzke, C., Kardjilov, N., Lenoir, N., Manke, I., Oswald, S.E., Tengattini, A. (2019). What comes NeXT? – High-speed neutron tomography at ILL. *Opt. Express*, 27, 28640.

Trtik, P. and Lehmann, E.H. (2016). Progress in high-resolution neutron imaging at the Paul Scherrer Institut – The neutron microscope project. *J. Phys. Conf. Ser.*, 746, 012004.

Villanova, J., Daudin, R., Lhuissier, P., Jauffrès, D., Lou, S., Martin, C.L., Labouré, S., Tucoulou, R., Martínez-Criado, G., Salvo, L. (2017). Fast in situ 3D nanoimaging: A new tool for dynamic characterization in materials science. *Mater. Today*, 20, 354–359.

Wildenschild, D. and Sheppard, A.P. (2013). X-ray imaging and analysis techniques for quantifying pore-scale structure and processes in subsurface porous medium systems. *Adv. Water Resour.*, 51, 217–246.

Part 2

Illustrations

6

Large-Scale Mineral Precipitation and Its Effects on Infrastructures

Anna RAMON-TARRAGONA and Eduardo E. ALONSO

Technical University of Catalonia, Barcelona, Spain

6.1. Introduction

A rock mass or a soil can swell if it contains soluble salts capable of being dissolved in water and precipitating as crystals in joints or voids of the rock or in soil porosity. The anhydritic and gypsiferous rocks and soils are affected by this mechanism. Expansion in these materials is explained by the precipitation of gypsum crystals in the rock discontinuities from supersaturated groundwater in calcium sulfate, or, in the case of compacted sulfated soils, by the crystallization of ettringite and thaumasite minerals when a sulfate attack develops within the compacted soil (Alonso et al. 2013; Alonso and Ramon 2013a, 2013b; Ramon and Alonso 2018; Mohamed 2000). This phenomenon leads to severe observed expansions and may damage infrastructures and buildings. However, the presence of gypsum or anhydrite in the rock mass or soil does not automatically lead to the development of swelling phenomena; additional conditions, described in this chapter, are required. The prediction of strains and swelling pressures in sulfated formations is a difficult task.

This chapter addresses the development and numerical modeling of anhydritic rock expansion induced by crystal growth and its interaction with infrastructures. In addition to a general view of the potential effects of expansions of sulfated rocks and

Salt Crystallization in Porous Media,
coordinated by Hannelore DERLUYN and Marc PRAT.

soils to infrastructures, the current understanding of the expansiveness observed in anhydritic formations will be described through the analysis of two case histories of infrastructures impaired by expansions: the cases of the Lilla tunnel and the Pont the Candí viaduct. They have been selected because the severity of expansions in the Lilla case stands out among the rest of the reported crystal-induced expansions in tunnels and the nature of heave of the viaduct of Pont de Candí is considered rather unique. Also, extensive instrumentation and field investigation data are available for both of them. In addition to highlighting the main features of the expanding mechanism, a numerical modeling capable of simulating rock expansions induced by crystallization and addressing infrastructure-anhydritic rock interaction will be described. Calculations will be compared with field records from the damaged infrastructures sites.

6.2. Damaging effects of expansions in sulfated rocks and soils

Tunnels excavated in natural formations containing anhydrite ($CaSO_4$) and gypsum ($CaSO_4 \cdot H_2O$) are the most common infrastructures damaged by the expanding phenomena of sulfated rock. The Triassic rocks of Anhydritgruppe (Middle Muschelkalk) and Gipskeuper (Middle Keuper) found in Baden-Württemberg (Germany) and Jura Mountains (Switzerland) have a high sulfates content and have been recurrent for decades when swelling problems appear in a tunnel excavation. Several authors describe case histories of tunnels affected by expansions crossing these formations in Central Europe (Grob 1976; Kovári et al. 1988; Anagnostou 1993, 2007; Steiner 1993, 2020; Madsen et al. 1995; Einstein 1996; Wittke-Gattermann and Wittke 2004; Wittke 2006, 2014; Berdugo 2007; Anagnostou et al. 2010, 2015; Alonso et al. 2013; Wittke et al. 2017; Ramon and Alonso 2018; Steiner and Schwalt 2019). Severe expansions also affect tunnels crossing Tertiary sulfated claystone formations (Alonso et al. 2013). The Wagenburg tunnel, in Germany, and the Lilla tunnel, in Spain, are two examples of them. Both anhydritic-gypsiferous formations have the presence of a highly mineralized groundwater in common, with sulfate and calcium as the main dissolved ions (Alonso et al. 2013).

These expansions generate large displacements on non-confined surfaces in tunnels and high swelling pressures against structural elements. In general, the intensity of swelling in formations with anhydrite and gypsum exceeds by far the expansivity recorded in expansive hard rocks and soils free of sulfates (Berdugo 2007) (Figure 6.1). Heave values close to 1 m have been measured in the flat slab of tunnels excavated in sulfated rocks and also swelling pressures in excess of 6 MPa have been recorded against tunnel lining in these formations. Two interesting characteristics of these phenomena are the rapid occurrence of expansions and their uncertain evolution over time.

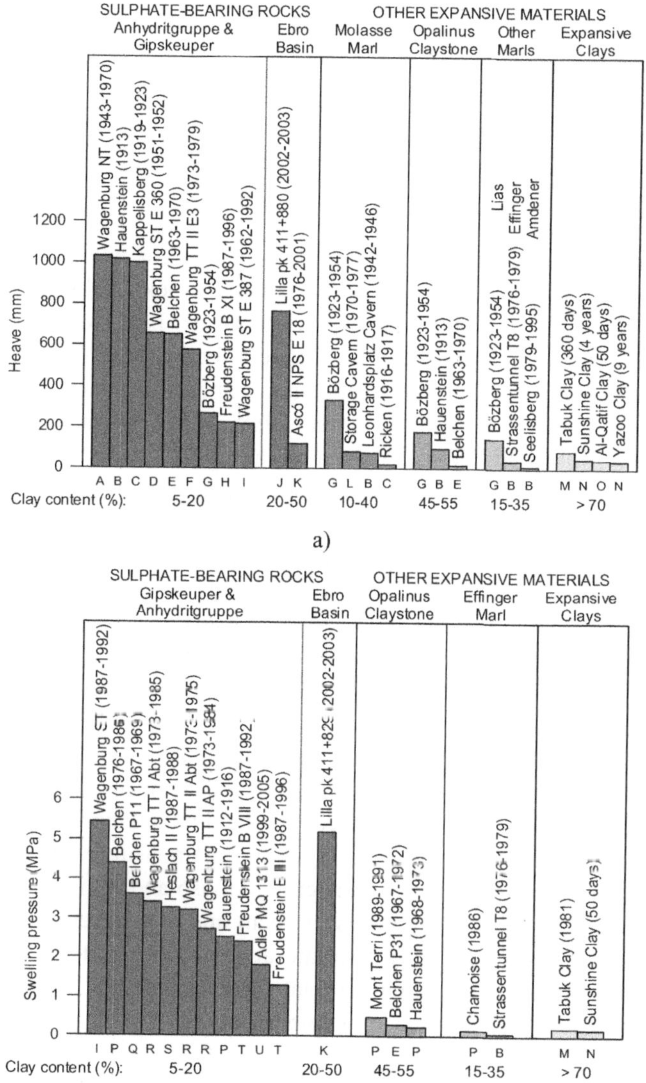

Figure 6.1. *Extreme expansive phenomena recorded in tunnels, caverns, deep excavations and foundations in swelling rocks and soils, compiled by Berdugo (2007)*

Swelling in anhydritic formations not only affects tunnels, it may affect a wide typology of infrastructures and buildings. This is the case of the significant heave of the Pont de Candí viaduct induced by the development of expansions below its piled foundations (Alonso and Ramon 2013; Ramon and Alonso 2013; Alonso et al. 2015), the damaging heaves of the historical center of Staufen (Germany) and houses in Lochviller (France) (Sass and Burbaum 2010; Ruch and Wirsing 2013) and the uplift of a public building in Spain built in an ancient quarry of gypsum (Ramon and Alonso 2018). The reported cases highlight the widespread occurrence of these phenomena.

Expansions involving sulfated materials can also occur within fills made up of compacted material from excavations in sulfated rocks. This is the case of expansions found when sulfate attacks cement- or lime-treated sulfated clay material. A loss of strength and a surface heave affecting road bases and sub-bases have often been reported when the treatment is applied to relatively thin layers of these materials (Sherwood 1962; Hunter 1988; Mitchell and Dermatas 1992; Snedker 1996; Mohamed 2000; Puppala et al. 2003; Rajasekaran 2005). This process is capable of inducing high swelling pressures against rigid structures and important heave when the attack massively affects layers of a certain thickness. An example is the phenomenal swelling of two railway embankments located in Spain (Alonso and Ramon 2013b). The scale of the reaction of sulfate attack in these embankments, which extended to thicknesses of 8–10 m, is considered to be quite unique. The massive growth of ettringite and thaumasite minerals in the embankments reinforced by Portland cement, due to the sulfate attack on cementitious materials, generated sustained important expansions. The induced swelling strains resulted not only in a surface heave but also in very high and totally unexpected forces against bridge abutments, which caused significant structural damage. Figure 6.2 shows the surface vertical displacements measured at the embankments and a scanning electron microscopy photograph of ettringite and thaumasite crystals found in one sample of the embankment expanding material. Alonso and Ramon (2013b) analyze the case of the expanding embankments in detail and also describe a simulation of the chemical reactions involved in a sulfate attack and a finite element model of embankment swelling. They calculate a total force against the bridge abutments of 2.32 MN/m induced by the expansion of embankments.

Knowledge of the fundamental mechanisms involved during the development of swelling, the soil and rock mass conditions leading to triggering the expansive phenomenon and the maximum pressure that could be exerted by expansions, helps for the prediction of the potential swelling development and the estimation of the intensity of the expansions. This is fundamental information for the proper design of safe tunnels, excavations and embankments in sulfated materials.

a) b)

Figure 6.2. *(a) Heave measured at the embankments surface; (b) ettringite and thaumasite crystals found in the expanding material (Alonso and Ramon 2013b). For a color version of this figure, see www.iste.co.uk/derluyn/salt.zip*

6.3. Extreme expansive phenomena in the Lilla tunnel

6.3.1. *Geology, design and construction*

The Lilla tunnel belongs to the high-speed railway line Madrid-Barcelona. It is located in the province of Tarragona, close to Montblanc city. The tunnel is excavated in the Lower Ebro basin, the largest evaporite deposit in Spain. The excavation entirely involves Eocene sulfated formations. The site of the tunnel experienced a very intense tectonic activity in the past. The Eocene claystone formation is folded and compressed by the thrust exerted by Triassic and Palaeozoic rocks which resulted in faults and folded strata.

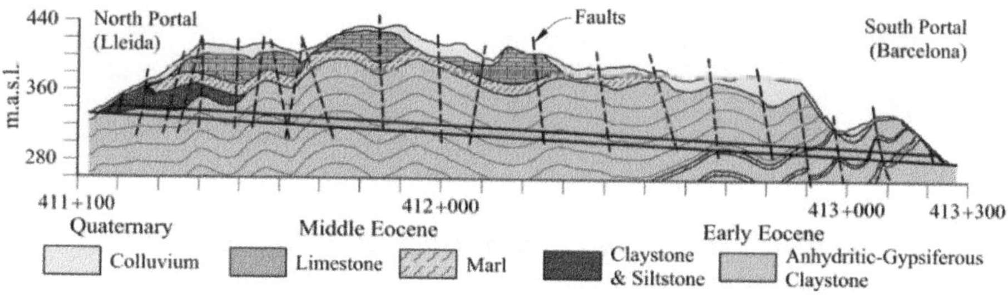

Figure 6.3. *Longitudinal geological profile of the Lilla tunnel. For a color version of this figure, see www.iste.co.uk/derluyn/salt.zip*

Figure 6.3 shows the longitudinal geological profile along the Lilla tunnel. The tunnel crosses early Eocene brown clay rocks containing anhydrite nodules and intricate cross-shaped fibrous gypsum veins (Figure 6.4(a)). A singular aspect of the excavated material is the existence of a persistent system of slickensided surfaces due to the previous intense tectonic activity occurred during the Bartonian uplift (Figure 6.4(b)).

The tunnel has an approximate length of 2 km and the overburden ranges from 32 to 110 m. It was initially excavated with a horse-shoe cross-section of 117 m^2 (Figure 6.5). No specific precautions were taken regarding swelling phenomena. The excavation of the tunnel was performed mainly by means of drill and blast from both portals following the methodology of the new Austrian tunneling method (NATM). The section was divided into top heading excavation and subsequent benching. The reason for using explosives during the excavation was the high strength of the claystone. Sprayed concrete was immediately applied systematically after excavation and rock bolts and steel arch ribs were installed to support zones of low rock quality. A 300-mm-thick cast-in-place unreinforced concrete built at the vault and abutments and a 300-mm-thick flat slab at floor level defined the final lining of the original horseshoe cross-section of the tunnel (Figure 6.4). The tunnel floor suffered the action of environmental agents (occasional wetting, periodic changes in relative humidity) and the circulation of heavy machinery during the major part of the construction process.

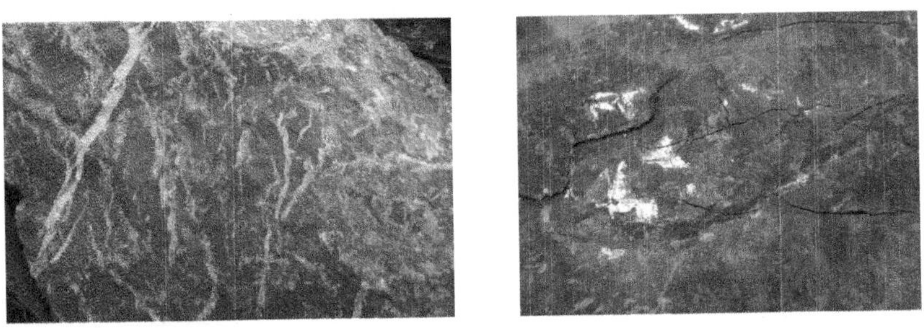

a) b)

Figure 6.4. *Details of the claystone formation. (a) Gypsum veins in the clay matrix; (b) slickenside. For a color version of this figure, see www.iste.co.uk/derluyn/salt.zip*

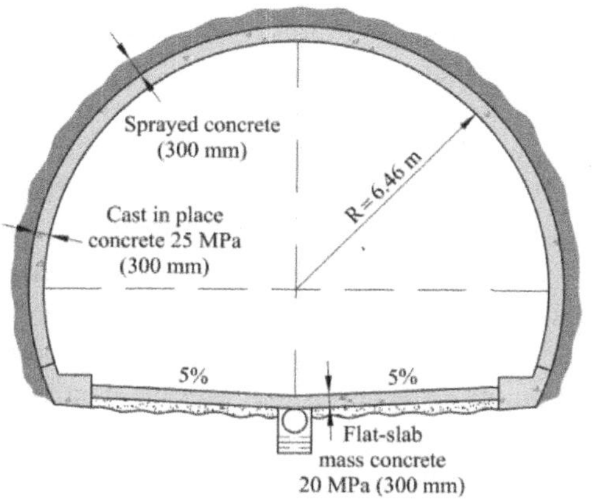

Figure 6.5. *Initial representative cross-section of the Lilla tunnel. For a color version of this figure, see www.iste.co.uk/derluyn/salt.zip*

6.3.2. *Expansive phenomena and ground properties*

Heave at floor level of the Lilla tunnel was first detected in September 2002, just after the construction of the flat slab, immediately after building the initial horseshoe cross-section. The severity of expansive phenomena increased very rapidly in time without any signs of stabilization (Alonso et al. 2013) and resulted in the local failure of the concrete slabs and the breakage of the longitudinal drainage system. Expansions were concentrated at the tunnel floor (Figure 6.6). The abutments and the crown were not affected by ground expansions. In fact, no signs of fissures in the shotcrete and the lining of the vault were observed.

a) b) c)

Figure 6.6. *Heave and failure of the flat slab: (a) March 2003; (b) May 2003; (c) September 2003. For a color version of this figure, see www.iste.co.uk/derluyn/salt.zip*

With the aim of investigating the origin of the heave of the floor slab and, as a first step toward the design of a cross-section able to cope with the observed swelling phenomena, a study of ground conditions based on "in situ" and laboratory investigation campaigns was set out. Some tunnel test sections with a flat slab and others, in which a provisional reinforced curved invert was built, were monitored by means of high-precision "sliding micrometers", installed along boreholes drilled from the tunnel floor, with the purpose of measuring the vertical displacements in depth. Pressure cells were installed at the concrete–claystone interface of the vault and invert to measure swelling pressures. Vibrating wire piezometers were also placed in the same monitoring sections, at varying depths, to investigate water levels. Continuous rock cores were recovered from the tunnel floor by means of boreholes 20 m long. The recovered material in boreholes was investigated in detail by means of mineralogical, chemical and mechanical laboratory tests.

The behavior of a circular cross-section was also investigated at a large scale. This required a maximum excavation of 4 m under the original flat slab of the tunnel. Three different types of support design – full resisting support and two different types of yielding support conditions (Kovári et al. 1988) – were built along three test cylindrical stretches inside the tunnel. They were tested at full scale. The worst expected swelling conditions were approximated during the tests by flooding the rock below the lower part of lining during 30 days and then the water was removed. Monitoring of sliding micrometers and pressure cells over 400 days showed the performance of the three circular sections.

X-ray diffraction analysis on the material recovered from boreholes indicated that the excavated material is dominated by a clayey matrix, made of non-expansive clay minerals, having a relative content ranging from 51% to 67% in weight (mainly illite and paligorskite). Quartz, dolomite and a crystalline fraction made of gypsum and anhydrite were also identified in the samples analyzed. A water level close to the tunnel floor was observed in drilled boreholes. Groundwater had a high sulfate content.

The monitoring data indicated that the magnitude of the expansive phenomenon in the Lilla tunnel was not homogeneous along the entire length of the tunnel. The extreme swelling concentrated along the first kilometer from the North portal and no swelling was detected along 500 m close to South portal. Figures 6.7 and 6.8 show that the expansive phenomena under the flat slab and invert evolved rapidly. In fact, a comparison among published field heave records of tunnels indicated that Lilla provided one of the fastest rates of development of floor heave and swelling pressures against the invert. Instrumented critical sections with flat slab showed a heave higher than 700 mm after 13 months of topographic measurements with a vertical displacement rate ranging from 1.2 to 2 mm/day. Pressure cells indicated that the radial pressures against the tunnel floor lining were very high (4–5 MPa) at some points of the invert lining, although relatively high vertical displacements were measured at the same locations (Figures 6.7 and 6.8).

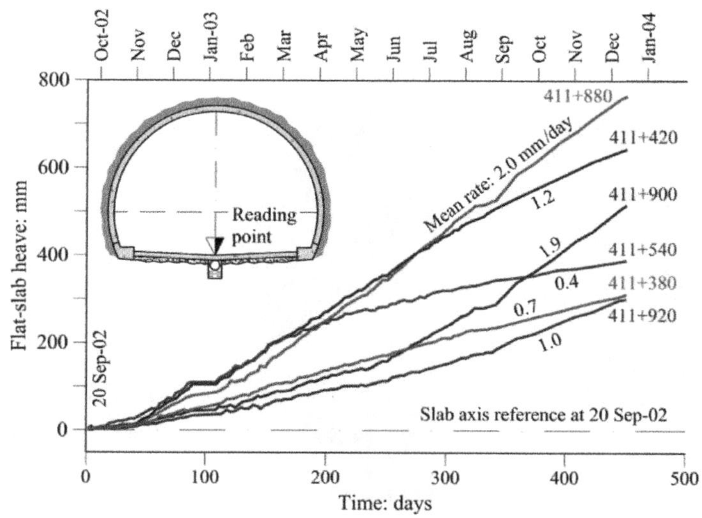

Figure 6.7. *Evolution of the floor heave in the indicated tunnel sections with flat-slab. For a color version of this figure, see www.iste.co.uk/derluyn/salt.zip*

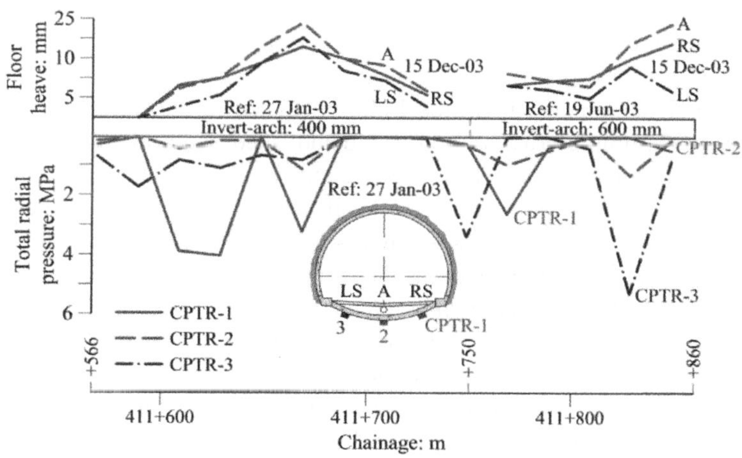

Figure 6.8. *Pressure against stress cells and heave measured in invert test sections. For a color version of this figure, see www.iste.co.uk/derluyn/salt.zip*

Vertical strains recorded by means of continuous extensometers indicated precisely that swelling deformations concentrated along a 4- to 5-m-thick stratum (active zone) located below the tunnel floor. The detailed study of the material recovered from the boreholes found the presence of thin needles of gypsum crystals and gypsum aggregates in "rosettes" (Figure 6.9) in several open discontinuities in the cores recovered from the active zone. On occasions, rosettes were found to cover rock discontinuities almost completely.

Figure 6.9. *(a) Gypsum needles and (b) planar gypsum "rosettes" found on claystone discontinuities of cores recovered in the Lilla tunnel. For a color version of this figure, see www.iste.co.uk/derluyn/salt.zip*

6.3.3. *Tunnel reinforcement and reconstruction*

The exceptional expansions in the Lilla tunnel led to the reconstruction of the entire tunnel. The initial cross-section was demolished and a circular cross-section was built along the tunnel. The cross-section was designed according to the "resistance principle" in view of the uncertainty of the long-term behavior of yielding supports. A highly reinforced high strength (80 MPa) concrete lining was built. This concrete strength far exceeds the strengths of the typical concretes used in tunnel linings. The design swelling pressures were selected from field measurements recorded at the instrumented test sections: 4.5 MPa for the invert and 2.5 MPa for the vault (Marí and Pérez 2003; Marí and Molins 2007; Alonso et al. 2013). The circular structure was calculated under two-dimensional, plane-strain conditions. The lining design relied on a non-uniform distribution of swelling pressures against the tube, following a conservative criterion to define the lining thickness and reinforcement. Alonso et al. (2013) describe the details of the procedure followed for the design of the tunnel lining.

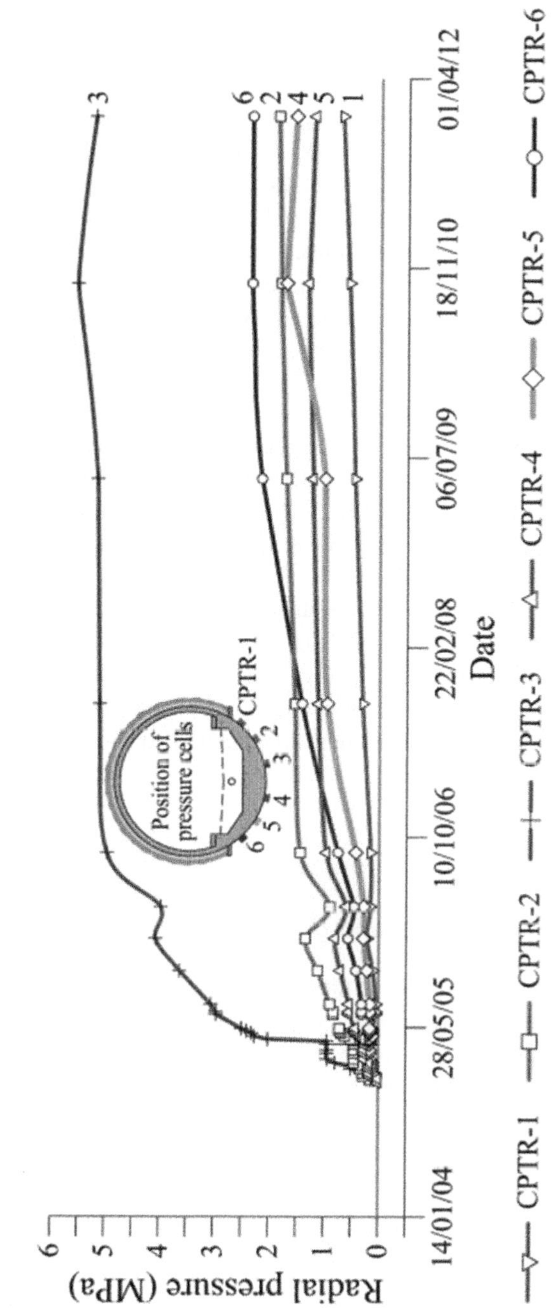

Figure 6.10. *Radial pressures recorded during tunnel operation in one section. For a color version of this figure, see www.iste.co.uk/derluyn/salt.zip*

The tunnel lining was instrumented by means of loading cells installed at the rock/lining interface and strain meters attached to the reinforcement steel bars to investigate the radial pressures against the lining and the stress of steel bars.

The maximum swelling pressures measured during the operational stage of the Lilla tunnel stays in the range from 5 to 6.7 MPa (January 2005 to December 2011). The records indicate that these values tend to be essentially stabilized. The evolution of the recorded radial pressure in one section is shown in Figure 6.10. In addition, the analysis of instrumentation records showed that the heavily reinforced high strength concrete circular lining works in compression despite the extremely high swelling pressures recorded in some positions (Alonso et al. 2022). The steel reinforcement is also under compressive stresses at virtually all measuring points. The magnitude of the stresses is small, very far from yielding conditions.

The most significant result of tunnel monitoring is the extreme variability of recorded radial pressures along the tunnel and also at a given cross-section: one pressure cell may register values in excess of 5 MPa, while the remaining cells in the same section, closely located, register very low or even no pressures (Figure 6.10).

The detailed analysis of recovered cores drilled from the ground surface from areas with different swelling intensity of the tunnel floor indicated that the conditions that favor the development of anhydritic-induced heave and pressure in the Lilla tunnel are as follows: a significant proportion of sulfate minerals (specially anhydrite), certain rock damage (presence of sheared discontinuities) and a significant presence of clay (Alonso et al. 2013).

6.4. Heave of the Pont de Candí Viaduct

6.4.1. *Performance of the viaduct and geology*

The Pont de Candí viaduct is located in the vicinity of the North Portal of the Lilla tunnel. It is one of the main bridge structures of the Madrid-Barcelona high-speed railway line in Spain. Figure 6.11 shows a view of the viaduct. The deck of the bridge has a length of 413 m and is supported by nine pillars with a maximum height of 56 m at the center of the valley. Each pillar is supported by a deep foundation constituted by a group of nine large bored piles, 1.65 m in diameter and 20 m long on average (Figure 6.12).

Figure 6.11. *The Pont de Candí viaduct. For a color version of this figure, see www.iste.co.uk/derluyn/salt.zip*

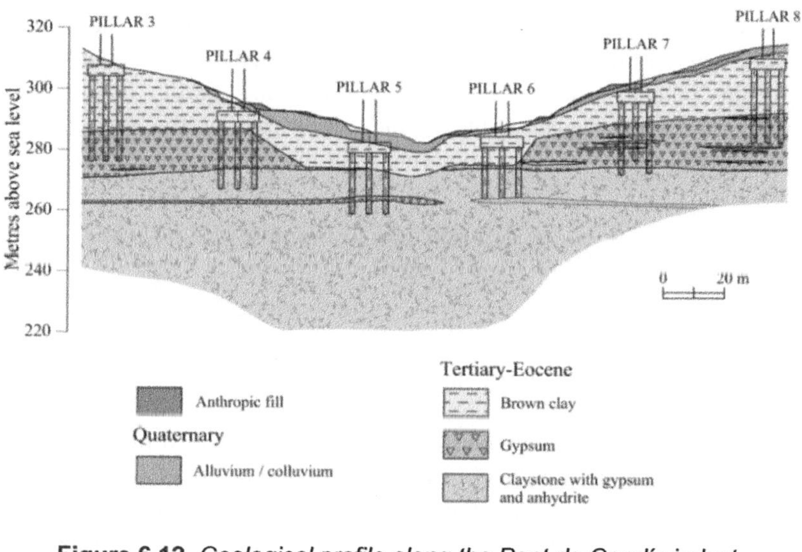

Figure 6.12. *Geological profile along the Pont de Candí viaduct. For a color version of this figure, see www.iste.co.uk/derluyn/salt.zip*

The valley crossed by the bridge is covered in its central part by a 5-m-thick deposit of soft alluvium and colluvium soils, which overlies a 15-m-thick layer of low-plasticity non-active brown clays and gypsum levels of Eocene age. Below, there is a harder red sulfated claystone formation, also of Eocene age, which was judged strong enough to support the large diameter piles of the central pillars of the bridge (P3 to P7). This stiff claystone is the same formation crossed by the Lilla tunnel. A continuous horizontal layer of gypsum (0.6–2 m thick) divides the red claystone unit into an upper and a lower level. The upper claystone level presents high gypsum content. Anhydrite was not detected in this upper level. In sharp contrast, the lower claystone unit below the thin continuous gypsum layer has high anhydrite content and increased strength (Figure 6.12). Water content clearly varies along the different layers. The hard claystone stratum has low water content; the values are smaller in the lower anhydritic claystone when compared to the ones in the upper level of this stratum. In addition, the water content increases rapidly when moving upwards within the upper brown clays, which are essentially saturated (Alonso and Ramon 2013). This is a significant result, which helps to define a scenario for the observed heave phenomena.

Periodic leveling of the structure, performed after the construction of the viaduct, detected the development of a progressive heave of the central pillars of the viaduct (Alonso and Ramon 2013). A sustained heave rate ranging from 5 to 10 mm/month accumulated without an apparent stabilization, summing up to a heave close to 25 cm at the center of the bridge at the beginning of investigations. In addition to the heave of the pillars, small rotations along the pillar shafts, longitudinal and transversal horizontal displacements at the upper part of the pillars, and also rotations and vertical and lateral displacements of the pile caps of the central pillars, were measured.

6.4.2. *Field investigation*

An extensive field investigation was carried out to understand the origin of the heave and the state of the entire structure (Figure 6.13).

Vertical boreholes with continuous core recovery were drilled from the surface of pile caps of the four central pillars of the viaduct at the center of twelve piles (three boreholes per pillar) to investigate the foundation integrity (Figure 6.13). Boreholes penetrated a few meters in the natural ground below the pile tips. Boreholes were inspected by means of a high-resolution acoustic televiewer, a multi-arm caliper and an optical video camera. The inspection of the walls of the boreholes revealed the systematic presence of tensile fissures at the interface between the pile's heads and the 3.5-m-thick pile cap. These fissures indicated that the piles had been subjected to significant and unexpected stresses. A semi-analytic analysis of a pile group swelling interaction, described below in section 6.4.4, explained that the cracks observed in the cap-pile contacts were a result of the development of heterogeneous expansions in depth, below the pile tips (Alonso et al. 2015).

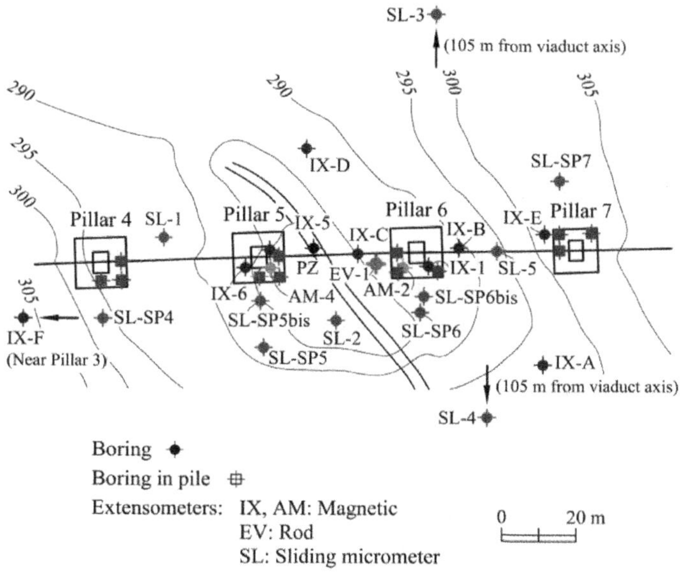

Figure 6.13. *Surveying boreholes and deep extensometers in the Pont de Candí viaduct area. For a color version of this figure, see www.iste.co.uk/derluyn/salt.zip*

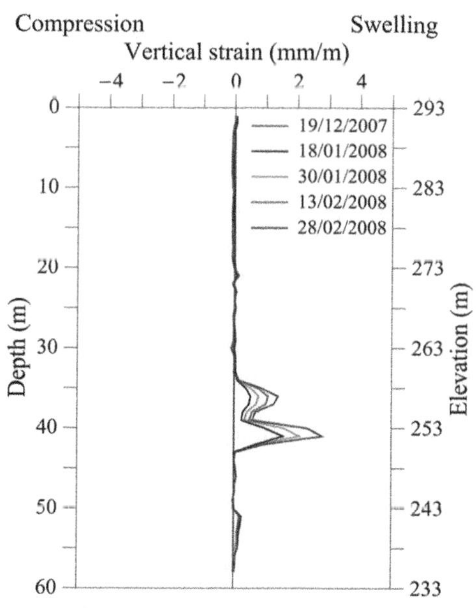

Figure 6.14. *Vertical strains recorded in long extensometer SL1. For a color version of this figure, see www.iste.co.uk/derluyn/salt.zip*

Long deep continuous extensometers installed in boreholes drilled in the vicinity of the central pillars (Figure 6.13) identified a horizontal active expanding layer with a thickness ranging from 12 to 15 m. The active layer was located below the pile tips of the central pillars of the bridge, entirely within the anhydritic claystone. Figure 6.14 shows the measurements of vertical strains (mm/m) in a long extensometer placed between pillars 4 and 5.

Quantitative X-ray diffraction analysis on samples recovered in borings, at different depths, provided vertical profiles of gypsum and anhydrite content. These plots could be compared with extensometer data. The comparison indicated that swelling strains were directly associated with the presence of anhydrite. No expansions were measured where only gypsum was present (Figure 6.15). This information is very valuable for the interpretation of the swelling mechanisms.

The field investigation also identified a phreatic level 5 m below the valley crossed by the viaduct and a deep water level in the upper gypsiferous claystone.

The deep expansive phenomenon also generated a heave of the natural ground surface. Vertical displacements at surface level were measured by means of a network of surface topographic marks that covered an area 200-m wide centered along the viaduct axis. In general, the most significant heave was measured in an area around the central pillars of the viaduct (Figure 6.16). The evolution of the measured heave at the ground surface level indicated that the heave phenomenon did not stabilize.

A campaign of hydraulic cross-hole tests was performed at different depths to investigate the structure, fissuring and connectivity of discontinuities in the active expanding layer. The tests were performed along three boreholes drilled in an area where important vertical displacements had been measured at the ground surface, near pillar 5. The hydraulic analysis of the results showed that the active expansive layer had a system of open horizontal discontinuities that were hydraulically connected (Ramon and Alonso 2013). No hydraulic connection could be established in the vertical direction.

Rock cores recovered from the boreholes presented an increase of the density of fissures at depths corresponding to the active layer. Moreover, the presence of gypsum crystal growth was often observed in some open discontinuities in the material recovered from depths corresponding to the active expanding layer: gypsum crystal needles in open discontinuities, partially clogging the open fractures of the claystone matrix, and thin and warped sheets of gypsum embedded in the rock matrix (Figure 6.17).

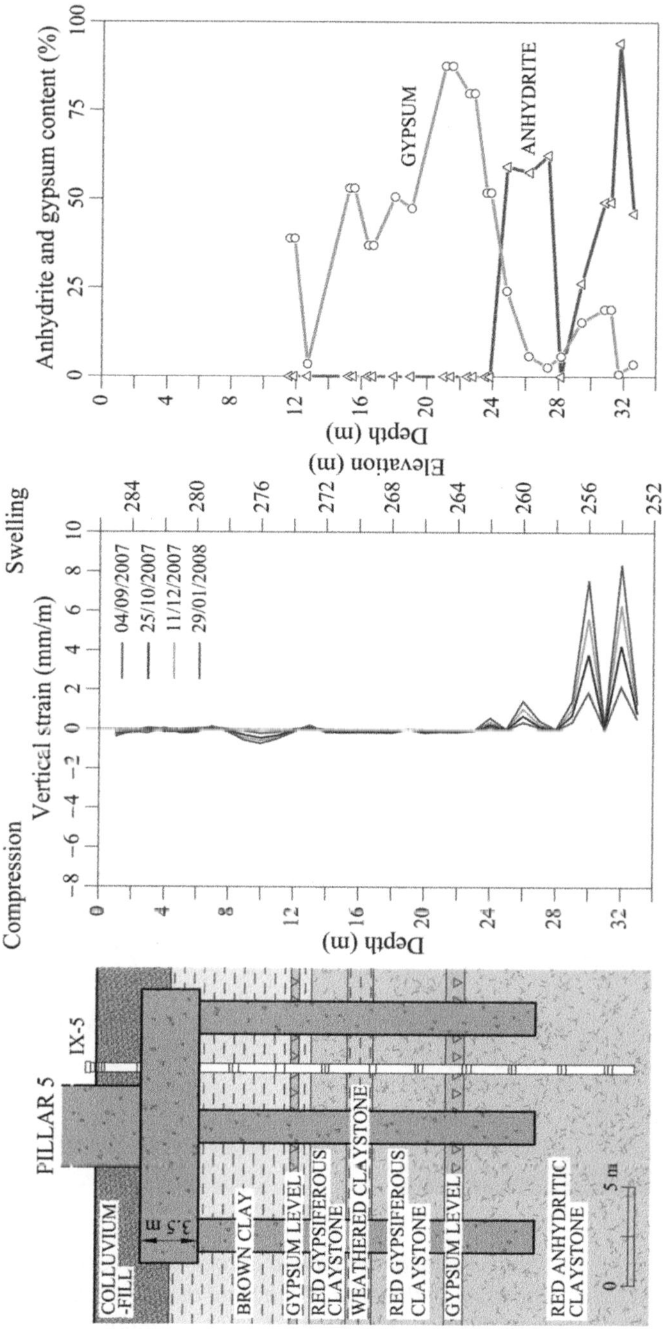

Figure 6.15. *Swelling strains and anhydrite and gypsum content in depth. For a color version of this figure, see www.iste.co.uk/derluyn/salt.zip*

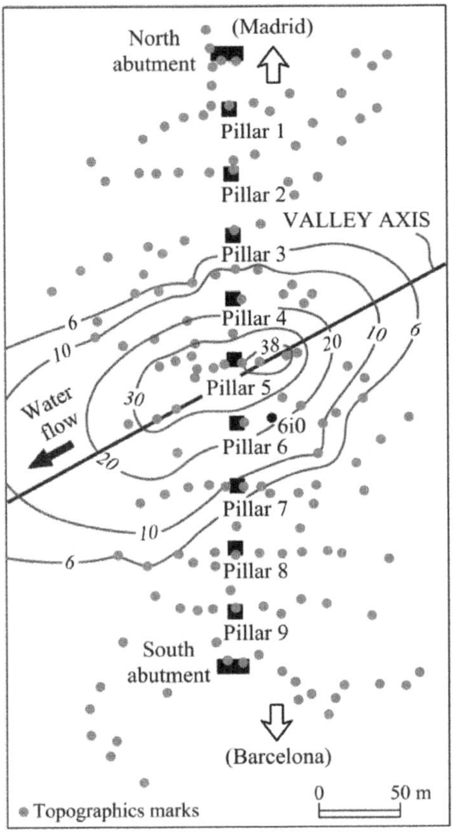

Figure 6.16. *Contours of equal surface heave (mm) (November 26, 2007 to April 30, 2008). For a color version of this figure, see www.iste.co.uk/derluyn/salt.zip*

Figure 6.17. *Gypsum crystal growth found in rock cores recovered from the expanding layer in the area of the Pont de Candí viaduct. For a color version of this figure, see www.iste.co.uk/derluyn/salt.zip*

6.4.3. *Remedial measures*

The progressive heave was initially compensated at the bridge deck level by shortening the bearing supporting structures located at the upper part of the pillars. However, there was a limit to this solution against the sustained heave. The definitive corrective measure applied was the construction of an embankment 33 m high, covering partially the valley of the viaduct. This remedial measure was based on the hypothesis that the application of vertical load on the active layer would reduce the heave rate and eventually, it would be able to stop it.

The rate of the viaduct heave decreased after the embankment construction as well as the rate of development of swelling strains at depth within the active layer, which was recorded by continuous extensometers.

6.4.4. *Reaction of a pile group against deep expansion*

Deep swelling in the area of the Pont de Candí viaduct motivated the development of a procedure to analyze the response of loaded piled foundations under the movements originating from the expansion of a certain volume of rock/soil mass with an arbitrary shape and location.

Alonso et al. (2015) approached the analysis by means of a semi-analytical procedure. The stages of the analysis performed are as follows: (i) estimation of the ground displacements under the stresses induced by pile loading, the external loads and the volumetric deformations imposed by the expansion; (ii) estimation of the deformation of the piles under external loading, and (iii) compatibility of both sets of displacements at a discrete number of points of the pile or at the pile-ground interface. The deformation of the pile structure is obtained by means of structural analysis. Displacements of the half space (the foundation soil) are calculated by a superposition criterion using three fundamental solutions for an elastic half-space: Boussinesq (1885) for point loading on the surface, Mindlin (1936) for concentrated forces inside the half space and Sagaseta (1987) for the displacements induced by the expansion or contraction of a spherical source. The case to be solved is sketched in Figure 6.18(a).

The procedure, once validated, has been applied to the pile group foundation of the central pillar of the Pont de Candí viaduct (Pillar 5). A first step was the estimation of the volumetric deep expansion from "in situ" measurements. Rotations and displacements experienced by the pile group have been well predicted. The analysis has shown that all the piles of a group develop considerable bending moments and tensile stresses, which explain the tensile cracks observed at the pile-cap contact. Figure 6.18(b) is a scheme of the deformation of the pile group.

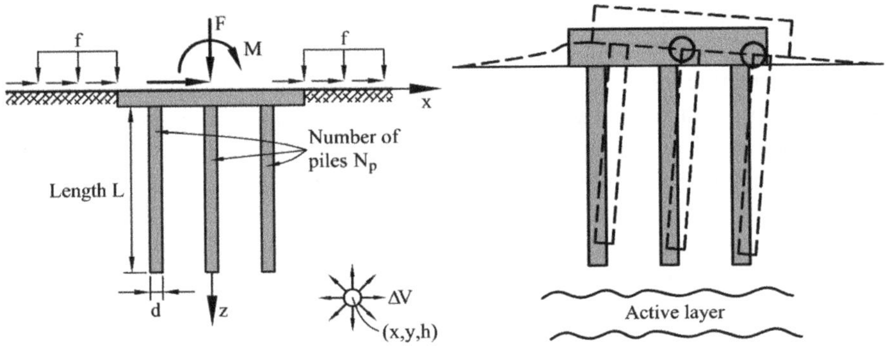

Figure 6.18. *(a) Problem description; (b) sketch of pile group deformation induced by expansion*

6.5. Swelling mechanisms in sulfated rocks

6.5.1. *Current interpretation*

Field observations made in cores recovered from the active expanding layers when interpreting the swelling experienced by the Lilla Tunnel, the heave at the area of the Pont de Candí viaduct and the uplift of a building close to Barcelona (Ramon and Alonso 2018), indicate that the origin of the observed swelling in the active layers is the precipitation of gypsum crystals in open discontinuities. In addition, field data indicate that the presence of anhydrite is required to trigger the expansion mechanism. A review of the records of tunnels affected by swelling crossing sulfated claystone in Central Europe supports this basic expansion mechanism.

Some laboratory tests have also shown evidence of crystal growth in rock fractures and that this phenomenon induces expansions in the rock. Gypsum crystal growth occurred at the surface of a powdered compacted sample after desiccation at the end of a free swelling test (Berdugo 2007). Free swelling tests on unconfined cores were also performed (Oldecop and Alonso 2012). During the free swelling tests, samples swelled and degraded. Gypsum crystals were found on existing crack surfaces at the end of the tests (Figure 6.19). The clay matrix, divided by the active cracks, became severely damaged and degraded.

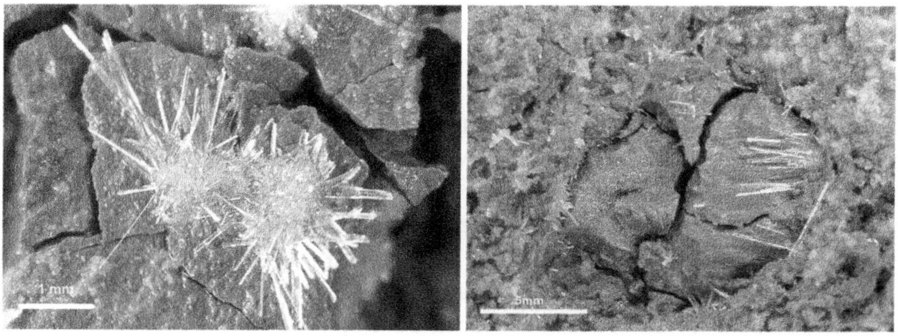

Figure 6.19. *Detail of gypsum crystal developing in rock discontinuities after tests. For a color version of this figure, see www.iste.co.uk/derluyn/salt.zip*

Gypsum crystals originate from the precipitation of calcium sulfate in water that is supersaturated in calcium sulfate with respect to gypsum. The experiments described by Kontrec et al. (2002) illustrate the kinetics of the transformation of anhydrite into gypsum in aqueous media. Scherer (1999), discussing the growth of crystals in pores, stresses that cracks and pits are sites which favor the nucleation of crystals. Charola et al. (2007), in their review of gypsum crystallization and its effect in the deterioration of natural stones, refer to observations of gypsum crystal growth in natural pores. They also mention the size of crystals as observed through microscopy and mention values in the range of microns.

Therefore, the process of precipitation of gypsum requires the existence of large enough fissures or voids, the presence of water and its supersaturation in sulfates. Two mechanisms capable to lead to supersaturated conditions of groundwater in the rock mass exist in connection with field observations in the Lilla tunnel and the Pont de Candí viaduct. The first one is the evaporation of groundwater and the second is the supersaturation conditions created by the dissolution of anhydrite.

The evaporation of water can occur at the rock-tunnel atmosphere interface. The relative humidity conditions imposed by the tunnel atmosphere, resulting from its ventilated environment, favors the evaporation toward the tunnel boundary at the exposed surfaces of the rock. However, a calculation on the amount of precipitated gypsum mass due to evaporation showed that the observed heave displacements of the planar floor slab required a mass of precipitated gypsum much higher than the mass that could be derived from water evaporation (Alonso and Olivella 2008). In addition, the evaporation of sulfated water cannot explain the swelling detected in depth, below the tips of the piles of the central pillars of the Pont de Candí viaduct.

The second mechanism is explained because the solubility of anhydrite ($CaSO_4$) is higher than the solubility of gypsum ($CaSO_4 \cdot H_2O$) at temperatures lower than 56°C. For a reference temperature of 15°C, water in contact with the anhydritic claystone will dissolve anhydrite until saturation conditions for calcium sulfate with respect to anhydrite are reached. Then, water will be supersaturated with respect to gypsum and the excess of calcium sulfate will tend to precipitate in gypsum crystals. This mechanism is potentially more intense and dangerous for engineering works than the evaporation-based mechanism. In addition, it is difficult to control, in contrast to the first mechanism. It can involve an arbitrary volume and an evaporation boundary is not necessary for its development.

The existence of fissures or voids in the rock favors the precipitation of crystals because they allow the flow of water through an impervious claystone and they provide the necessary space for the precipitation of crystals (Ramon and Alonso 2018). In addition, the measured pores of anhydritic claystone in the Lilla tunnel, which are in the range of a few nanometers (Pineda et al. 2014), provide an extremely reduced space for crystal precipitation and water flow is essentially restrained through them. But joints and slickensides provide a more favorable environment for crystal growth and flow rate. The observation that gypsum crystals started to grow in pre-existing discontinuities of the rock in crystallization experiments, reported by Oldecop and Alonso (2012), supports these ideas.

Therefore, there is certain evidence that in the materials investigated here gypsum crystals may precipitate in fractures. Supersaturation is achieved, in practice, when water flowing through discontinuities dissolves anhydrite (Figure 6.20). The process of crystal precipitation in discontinuities is thought to act as a local hydraulic "jack" that pushes apart the rock mass, resulting in the aperture of discontinuities, inducing swelling strains and pressures. Expansion will be controlled by the total mass of precipitated gypsum, the geometry of precipitated crystals and its interaction with the surrounding rock. Apparently, this process would be maintained while crystallization of gypsum occurs, which in turn is related to the supply of sulfated water.

The deformations and the high swelling pressures developed in the sulfated rocks are not directly related to the swelling processes associated with the hydration of clay minerals and it is well accepted that precipitation of crystals in fractures is at the origin of the swelling phenomena in expansive sulfated claystones. However, the presence of clay is thought to play a certain role in the expansive process (Ramon and Alonso 2018).

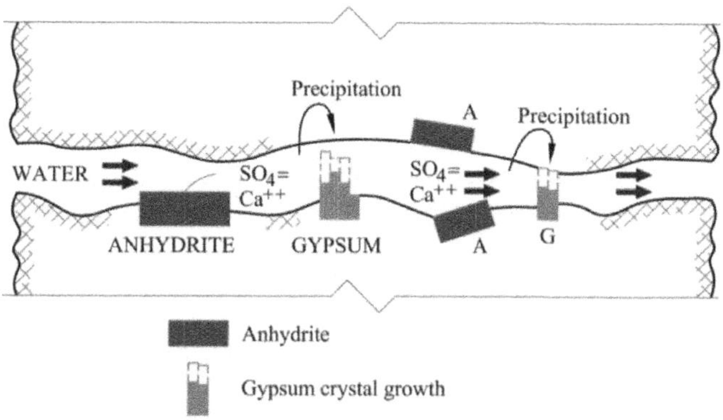

Figure 6.20. *Conceptual representation of gypsum precipitation on discontinuities (Ramon and Alonso 2018). For a color version of this figure, see www.iste.co.uk/derluyn/salt.zip*

6.5.2. *Crystallization pressures in anhydritic rocks*

The growth of a crystal from a supersaturated aqueous solution takes place at the interface between the crystal and the liquid (Scherer 1999). The accumulation of the crystalline material at the interface depends on the current concentration of solute, the solubility of the mineral and, also, on the pressure acting on the crystal.

Ramon and Alonso (2014) analyze the development of swelling pressures in pores and fractures of the rock due to crystal growth to estimate the possible maximum swelling pressure generated in sulfated rocks. They solve the problem by means of two complementary approximations: (a) the thermodynamic requirements for the equilibrium of the chemical reaction of precipitation or dissolution of a crystal immersed in an aqueous solution, and (b) the consideration of the superficial energy, γ_{cl}, at the interface between the crystal and the aqueous solution. The analysis that they performed is based on the approach described by Scherer (1999).

When a crystal approaches a solid surface (the wall of a pore or crack), a liquid film of solution remains at the wall-crystal interface. Scherer (1999) reports that the thickness of this film is a few nanometers. The film allows the growth of the crystal depending on its concentration. However, the solid wall will oppose crystal growth by repulsive forces. Ramon and Alonso (2014) found the crystal pressure against the walls of a rock pore considering three simple pore geometries based on field observations: a crystal growing in a cylindrical, planar and spherical pore. They suggest that the expressions found for the crystal pressure are an upper bound for pressures likely to develop “in situ”. The theoretical pressure derived for cylindrical pores (8.6 MPa) is in reasonable agreement with the maximum recorded swelling pressure against the Lilla tunnel (6.7 MPa). The mentioned upper bound limit is justified because gypsum crystal growth is a local phenomenon that takes place on fractures and the field swelling pressures measured at the rock-lining interface depend on the spatial heterogeneity of swelling sources.

6.5.3. *Scenarios of expansion and heave in anhydritic rocks*

The development of the swelling phenomena in anhydritic rocks is also controlled by the nature of construction works. The scenarios of expansion in the Lilla tunnel and heave of Pont de Candí viaduct, as well as the heaves of a building close to Barcelona (Ramon and Alonso 2018) and in the villages of Lochviller and Staufen (Sass and Burbaum 2010), clearly demonstrate that construction procedures have an effect on the triggering and development of crystal-induced expansions in these formations.

In the Lilla tunnel, the stress release under the horizontal tunnel floor, due to the excavation, induced the opening of the existing slickensided surfaces and favored also the development of new discontinuities in the rock. The presence of the tunnel, which acts as a drain, modified the groundwater regime in the vicinity of the tunnel. Sulfhate-rich water flow through discontinuities under the tunnel invert allowed the precipitation of crystals in fissures.

In the Pont de Candí viaduct, the hydraulic cross-hole tests performed demonstrated the existence of hydraulically connected horizontal fractures within the active layer and confirmed the presence of water in the expanding layer. It is thought that boreholes drilled at the design stage and, later, the construction of piles connected an upper aquifer with the deep fractured claystone. The entrance of water in the horizontal open fractures triggered the expansive phenomena.

6.6. Thermo-hydro-mechanical and chemical coupled modeling of crystal precipitation and rock deformation against infrastructures

Ramon and Alonso (2013) and Ramon et al. (2017) describe a numerical model that reproduces the observed swelling phenomena in anhydritic rocks and address the anhydritic rock–infrastructure interaction. It incorporates the occurrence of rock damage during the performance of excavations (i.e. tunnels and foundations), the crystallization of gypsum in discontinuities and a relationship between the precipitated mass of gypsum and the swelling strains developed, that depends on stresses in the rock mass. The model includes hydro-mechanical coupling and the transport of sulfate salts dissolved in the groundwater.

This section summarizes some details of the model formulation mentioned before and presents the simulation of the heave observed in the Pont de Candí viaduct and the expansions in the Lilla tunnel, and a sensitivity analysis to provide a deeper insight into the phenomena and its modeling. The simulations show the high capabilities of the numerical model to reproduce the rock expansions, heave and pressures recorded in the field. They illustrate the potential applications of the developed tool not only for the coupled analysis of rock expansion but also for the analysis and design of infrastructures in sulfated claystones.

Alonso et al. (2023) describe an alternative procedure to analyze and design the structures of tunnels built in anhydritic expanding rock from a different perspective. The methodology is very innovative and considers the heterogeneity of the expanding phenomena in the rock mass. The approach to analyze the effect of heterogeneous discrete volume expansion in rock proposed by Alonso et al. (2015) and described in section 6.4.4 can be applied to study the local structural response of piled foundations.

6.6.1. *Model formulation*

The swelling phenomenon is formulated within a general framework for coupled thermo-hydro-mechanical analysis for water saturated and unsaturated porous materials. The formulation of the specific problem of swelling due to gypsum crystal growth takes into account the presence of soluble gypsum and anhydrite, and the insoluble clay minerals in the mass balance of the solid phase (Ramon and Alonso 2013; Ramon et al. 2017). The mass balance of the solid phase (equation [6.1]) includes the basic mechanisms of dissolution and precipitation of gypsum and anhydrite, provides the variation of porosity, ϕ, in time and describes the volumetric strain rate induced by solid displacements of the porous medium, $\boldsymbol{u}$, and the volumetric strain rate induced by the precipitation or dissolution of crystals. The terms m_{gyp}, m_{anh}, ρ_{gyp} and ρ_{anh} in equation [6.1] are the precipitated or dissolved masses and densities of gypsum and anhydrite.

$$\frac{D_s\phi}{Dt}=(1-\phi)\nabla\cdot\left(\frac{d\boldsymbol{u}}{dt}\right)-\frac{1}{\rho_{gyp}}\frac{dm_{gyp}}{dt}-\frac{1}{\rho_{anh}}\frac{dm_{anh}}{dt} \qquad [6.1]$$

Two mass rate equations, depending on the “degree” of supersaturation or undersaturation of the solution in calcium sulfate, describe the kinetics of precipitation/dissolution of gypsum and anhydrite. Equation [6.2] refers to the kinetics of gypsum, a similar one is used for anhydrite. ω_l^m in equation [6.2] is the mass fraction of dissolved sulfate in water and $\omega_{l_{sat,gyp}}^m$ accounts for the equilibrium mass fraction in the presence of gypsum. κ is a constant describing the rate of dissolution/precipitation of mass of gypsum (kg/s m^2). The two equations take into account the specific surface of crystals in contact with water, σ_c (m^2 of crystal surface/m^3 of medium). In addition, the effects of temperature and pressure acting on crystals are considered for the calculation of the chemical equilibrium conditions of sulfated water ($\omega^m_{l_{sat,gyp}}$) according to Scherer (1999). The term ξ_{gyp} provides a sign to the mass rate of precipitation or dissolution.

$$\frac{dm_{gyp}}{dt} = \sigma_c \kappa \xi_{gyp} \phi_{gyp} \left(\left| \left(\frac{\omega^m_l}{\omega^m_{l_{sat,gyp}}(T,p)} \right)^{\theta} - 1 \right| \right)^{\eta} \qquad [6.2]$$

where

$$\xi_{gyp} = \frac{\omega^m_l - \omega^m_{l_{sat,gyp}}}{\left| \omega^m_l - \omega^m_{l_{sat,gyp}} \right|} \qquad [6.3]$$

The formulation of the model also describes the hydraulic and mechanical effects induced by precipitation. In particular, the mechanical effects of gypsum precipitation are taken into account in the momentum balance of the medium. The strains induced by precipitation of gypsum are calculated from the amount of precipitated volume of gypsum and the effect of prevailing stress acting on crystals (equation [6.4]).

$$\frac{d\varepsilon_i}{dt} = \frac{\gamma_i}{\rho_{gyp}} \frac{dm_{gyp}}{dt} \qquad [6.4]$$

where i = 1 defines the vertical strain rate component, and i = 2, 3 the horizontal ones. A given volume of precipitated crystals is expected to induce higher swelling strains under a lower confining stress. The γ parameter in equation [6.4] accounts for this effect and it is calculated from the prevailing stresses following an exponential function.

One additional equation for the mass conservation of solute is formulated for the transport of calcium sulfate solute to keep track of the minerals dissolved in water (equation [6.5]). In this equation, the precipitation and dissolution of sulfated minerals are treated as a source or sink of calcium sulfate solute and advective and diffusive mass flow rates are taken into account.

$$\rho_l \omega_l^m \nabla \cdot \left(\frac{du}{dt} \right) + \phi \frac{D_s \left(\rho_l \omega_l^m \right)}{Dt} + \nabla \cdot \left(\rho_l \omega_l^m q_l - \boldsymbol{D} \nabla \cdot \omega_l^m \right) = \\ = - \frac{dm_{gyp}}{dt} \left(1 - \frac{\rho_l \omega_l^m}{\rho_{gyp}} \right) - \frac{dm_{anh}}{dt} \left(1 - \frac{\rho_l \omega_l^m}{\rho_{anh}} \right) \qquad [6.5]$$

The formulation is completed by the description of rock damage, which is considered as a development of fractures that increases permeability of the rock and allows gypsum crystal growth. Fracture opening is calculated from calculated tensile deformations, usually associated with stress release induced by excavations (Ramon et al. 2017).

These formulations are included in the finite element program for thermo-hydro-mechanical analysis of porous media CODE BRIGHT (Olivella et al. 1996).

6.6.2. *Expansion calculation: comparison with field observations*

The thermo-hydro-mechanical and chemical (THMC) model described above was used to simulate a column of the rock mass of the foundation material of the central pillar of the bridge under plain strain conditions (Ramon and Alonso 2013). They represented the fractured and high permeable sulfated claystone as a porous material with a high porosity without fractures. The positions of expanding and stable layers are defined in the finite element geometrical model according to the records of expansions in several continuous extensometers. The horizontal flow conditions identified by the hydraulic cross-hole tests are reproduced in the active expanding layer.

Ramon and Alonso (2013) describe in detail the hypothesis made for the simulation and the analysis of the results. Figure 6.21 shows that the calculated vertical displacements reproduce the “in situ” measurements of heave at ground surface during the simulated 4-year period. The model reacts to the embankment construction and the calculated vertical displacement reproduce also the field measurements recorded after the embankment construction ($t = 924$ days in Figure 6.21).

The modeling of the damaging swelling behavior in the Lilla tunnel performed by Ramon et al. (2017) completed the validation of the general THMC expanding fracture model. Their finite element simulation analyses a cross-section of the Lilla tunnel located in the area affected by extreme expansions. The simulated tunnel cross-section has an overburden of 77 m and is modeled under plain strain conditions. The numerical analysis reproduces the process of tunnel excavation and the behavior of the initial flat-slab cross-section and the test section with an invert arch. The simulation involves also the construction of a concrete lining installation, including the invert arch.

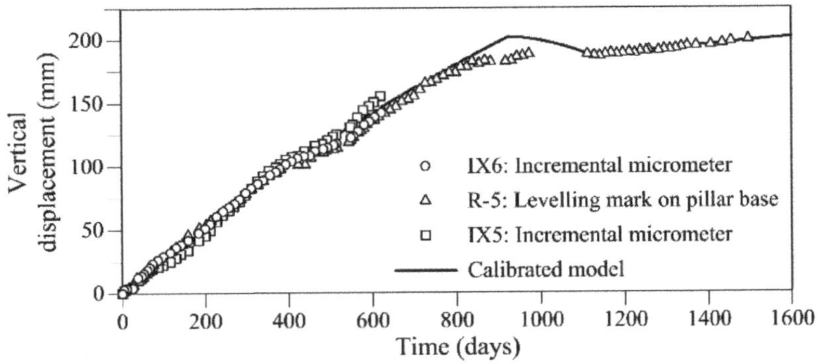

Figure 6.21. *Field measurements and calculated heave at the ground surface*

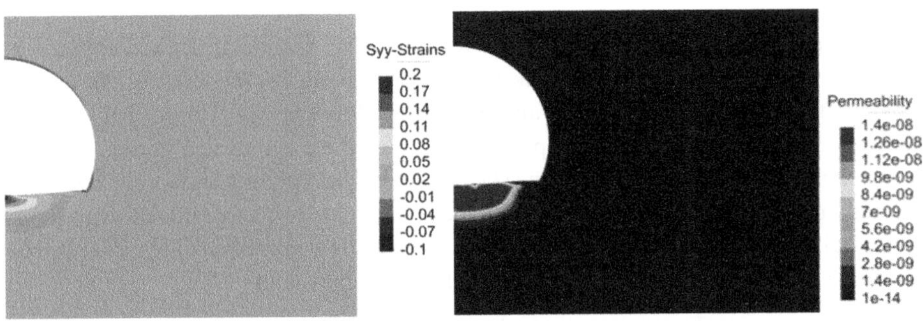

Figure 6.22. *Distribution of calculated (a) strains and (b) permeability (m²) after 1 year of gypsum precipitation. For a color version of this figure, see www.iste.co.uk/derluyn/salt.zip*

The predictions are in a reasonable good agreement with the records of floor heave, rock expansion and pressures against the invert. Figure 6.22 shows the distribution of strains and permeability calculated after 1 year of gypsum crystallization. The model allowed the analysis of the increase of fracture opening and permeability in the rock mass due to excavation and gypsum crystallization. Ramon et al. (2017) discuss its effects on the evolution of the rate of crystallization-induced expansions measured in time and on the progression of calculated gypsum and anhydrite transformations at depth. They also highlight that field heterogeneity and the development of rock degradation may explain some of the discrepancies among calculated and measured heave at increasing times.

The simulations served to analyze the influence of the structural reaction provided by the abutments and the invert of the tunnel on the expanding phenomena. A bulging shape of the flat surface of the tunnel floor, which is induced by the confining effect of the tunnel abutments is calculated, similar to field observations. Calculations also capture the development of pressures with reasonable precision under the invert and abutments.

6.6.3. *Sensitivity analysis*

Ramon and Alonso (2013) performed a sensitivity analysis considering the model used for the simulation of the Pont de Candí viaduct heave described above as a base case. Their analysis confirmed that the initial mineral content has an important effect on the development of heave; actually it controls the rates of dissolution and precipitation. In particular, they found that the anhydrite content has a stronger effect than the gypsum content on the calculated expansions. The study also shows that the equilibrium value of the concentration at saturation conditions when no stress is applied, which depends on the temperature, also has a significant effect on the vertical strains. They remark that the effects of stresses are substantial in the developed model. In contrast, the investigation shows that the initial porosity, the permeability and the gradient of induced flow have a small effect on the calculated heave.

6.7. Concluding remarks

Important expansions induced by large-scale precipitation of crystals can occur in the presence of anhydritic clayey rocks and, also, compacted soils built from excavated material in these geological formations. The cases presented illustrate that crystal-induced expansions are capable of severe and long-term damage to infrastructures. In addition, they show the main phenomenological aspects of the swelling observed when crystal precipitation occurs in sulfated materials and highlight that the chemical composition and the damage of the materials involved play a relevant role in the expansive response.

The mechanism that explains the expansions in sulfated rock formations is based on reaching supersaturated conditions in calcium sulfate dissolved in groundwater. Dissolution of anhydrite results in supersaturated conditions with respect to gypsum since the solubility of anhydrite is higher than the solubility of gypsum. This is believed to be a mechanism leading to large expansions. The presence of fractures in the rock favors the precipitation of gypsum when water and anhydrite are present.

A coupled fracture THMC model has been presented for simulating the expansions in sulfated rocks due to gypsum crystallization at a large scale. The formulation considers the presence of soluble sulfated minerals, the development of rock damage, the occurrence of precipitation and dissolution of crystals in fractures and the strains induced by crystal precipitation. A sensitivity analysis indicated that the initial mineral content, the solubility of gypsum and anhydrite and, especially, the confining stress have a relevant effect on the swelling strains calculated. The detailed campaigns of in situ investigations in the Pont de Candí viaduct and the Lilla tunnel served to verify the capabilities of the coupled model. The plain strain simulation of gypsum precipitation and expansions has been successful in both cases. Modeling reproduced the general characteristics of the expanding behavior observed; however, the field heterogeneity poses difficulties to the complete fitting of measurements along the entire length of the Lilla tunnel. Two additional calculation procedures have been also proposed for the analysis and design of infrastructures threatened by rock expansion.

Cases of compacted sulfated soils treated with cement or lime, in which expansions originate from ettringite and thaumasite crystallization developed during the progress of sulfate attack, were also mentioned in the chapter.

6.8. Acknowledgments

The authors thank the Spanish agency for railway infrastructure (ADIF) for the main financial and technical support during the development of an important part of the work reported in the chapter. The authors are also grateful to Dr. E. Tauler, Dr. M. Labrador and Dr. S. Galí from the Department of Crystallography of the University of Barcelona; Dr. C. Ayora from the J. Almera Institute; Dr. M. Saaltink from hydrogeological group research of Universitat Politècnica de Catalunya (UPC); Prof. E. Vázquez and Dr. M. Barrera from the Department of Building Engineering of UPC; and Dr. J. M. García and Dr. A. van Driessche from the Laboratory of Crystallographic Studies of the University of Granada/CSIC for the conceptual discussions on the nature of the observed expansions and heave and the conditions for crystal growth.

The Grant RTI2018-094226-J-I00 funded by the Spanish MCIN/AEI/ 10.13039/501100011033 and by "ERDF (FEDER) A way of making Europe", and, also, former financial support from the Spanish Public Works Ministry are gratefully acknowledged.

6.9. References

Alonso, E.E. and Olivella, S. (2008). Modelling tunnel performance in expansive gypsum claystone. In *Proceedings of the 12th Conference of International Association for Computer Methods and Advances in Geomechanics (IACMAG)*, Jadhav, M.N (ed.), Red Hook, New York.

Alonso, E.E. and Ramon, A. (2013a). Heave of a railway bridge induced by gypsum crystal growth: Field observations. *Géotechnique*, 63(9), 707–719.

Alonso, E.E. and Ramon, A. (2013b). Massive sulfate attack to cement-treated railway embankments. *Géotechnique*, 63(10), 857–870.

Alonso, E.E., Berdugo, I.R., Ramon, A. (2013). Extreme expansive phenomena in anhydritic-gypsiferous claystone: The case of Lilla tunnel. *Géotechnique*, 63(7), 584–612.

Alonso, E.E., Sauter, S., Ramon, A. (2015). Pile groups under deep expansion. A case history. *Canadian Geotechnical Journal*, 52(8), 1111–1121.

Alonso, E., Ramon-Tarragona, A., Verda, L. (2023). Designing tunnel lining in anhydritic claystones. Intensity and distribution of swelling forces. *Rock Mechanics and Rock Engineering*, 56(2), 1467–1487.

Anagnostou, G. (1993). A model for swelling rock in tunnelling. *Rock Mech. Rock Engng.*, 26(4), 307–331.

Anagnostou, G. (2007). Design uncertainties in tunnelling through anhydritic swelling rocks. *Felsbau*, 27(4), 48–54.

Anagnostou, G., Pimentel, E., Serafeimidis, K. (2010). Swelling of sulphatic claystones – Some fundamental questions and their practical relevance. *Geomech. and Tunnelling*, 3(5), 567–572.

Anagnostou, G., Serafeimidis, K., Vrakas, A. (2015). On the occurrence of anhydrite in the sulphatic claystones of the gypsum Keuper. *Rock Mech. Rock Engng.*, 48(1), 1–13.

Berdugo, I.R. (2007). Tunnelling in sulphate bearing rocks: Expansive phenomena. PhD Thesis, UPC, Barcelona.

Boussinesq, J. (1885). *Applications des potentiels à l'etude de l'equilibre et du mouvement des solides élastiques*. Gauthier-Villard, Paris.

Charola, A.E., Pühringer, J., Steiger, M. (2007). Gypsum: A review of its role in the deterioration of building materials. *Environ. Geol.*, 52, 339–352.

Einstein, H.H. (1996). Tunneling in difficult ground: Swelling behaviour and identification of swelling rocks. *Rock Mech. Rock Engng.*, 29(3), 113–124.

Grob, H. (1976). Swelling and heave in Swiss tunnels. *Bull. – Assoc. Engng. Geol.*, 13, 55–60.

Hunter, D. (1988). Lime-induced heave in sulphate-bearing clay soils. *J. Geotech. Eng., ASCE*, 114, 150–167.

Kontrec, J., Kralj, D., Brečević, L. (2002). Transformation of anhydrous calcium sulphate into calcium sulphate dihydrate in aqueous solutions. *Journal of Crystal Growth*, 240(1–2), 203–211.

Kovári, K., Amstad, C., Anagnostou, G. (1988). Design/construction methods: Tunnelling in swelling rocks. In *Key Questions in Rock Mechanics: Proceedings of the 29th US Symposium Rock Mech. Minnesota*, Cundall, P.A., Sterling, R.L., Starfield, A.M. (eds). Rotterdam.

Madsen, F., Flückiger, A., Hauber, L., Jordan, P., Vögtli, B. (1995). New investigations on swelling rocks in the Belchen tunnel, Switzerland. *Proc 8th Int. Congress on Rock Mechanics, Gifu*, 1, 263–267.

Marí, A.R. and Molins, C. (2007). Design of the strengthening of tunnel linings under localized pressures. In *Underground Space – The 4th Dimension of Metropolises*, Barták, J., Hrdina, I., Romancov, G., Zlámal, J. (eds). Taylor & Francis Group, London.

Marí, A.R. and Pérez, G.A. (2003). Refuerzo del revestimiento de los túneles de Lilla y Camp Magré para la Línea de Alta Velocidad Madrid-Zaragoza-Barcelona-Frontera Francesa. Tramo Lleida-Martorell. Provincia de Tarragona. UPC Internal Report.

Mindlin, R.D. (1936). Force at a point in the interior of a semi-infinite solid. *Journal of Applied Physics*, 7, 195–202.

Mitchell, J.K. and Dermatas, D. (1992). Clay soil heave caused by lime-sulphate reactions. In *Innovations and Uses for Lime*, Walker Jr., D.D., Hardy, T.B., Hoffman, D.C., Stanley, D.D. (eds). ASTM STP 1135. ATM Int., West Conshohocken, PA.

Mohamed, A.M.O. (2000). The role of clay minerals in marly soils on its stability. *Engineering Geology*, 57(3–4), 193–203.

Oldecop, L. and Alonso, E.E. (2012). Modelling the degradation and swelling of clayey rocks bearing calcium-sulphate. *Int. J. Rock Mech. Min. Sci.*, 54, 90–102.

Olivella, S., Gens, A., Carrera, J., Alonso, E.E. (1996). Numerical formulation for a simulator (CODE_BRIGHT) for the coupled analysis of saline media. *Engng. Comput.*, 13, 87–112.

Pineda, J.A., Alonso, E.E., Romero, E. (2014). Environmental degradation of claystones. *Géotechnique*, 64(1), 64–82.

Puppala, A.J., Wattanasanticharoen, E., Punthutaecha, K. (2003). Experimental evaluations of stabilisation methods for sulphate-rich expansive soils. *Ground Improv.*, 7(1), 25–35.

Rajasekaran, G. (2005). Sulphate attack and ettringite formation in the lime and cement stabilized marine clays. *Ocean Engineering*, 32, 1133–1159.

Ramon, A. and Alonso, E.E. (2013). Heave of a railway bridge: Modelling gypsum crystal growth. *Géotechnique*, 63(9), 720–732.

Ramon, A. and Alonso, E.E. (2014). Crystal growth and soil expansion: The role of interfacial pressure and pore structure. In *Unsaturated Soils: Research & Applications (UNSAT 2014)*, Khalili, N., Russell, A., Khoshghalb, A. (eds). CRC Press, Sydney.

Ramon, A. and Alonso, E.E. (2018). Heave of a building induced by swelling of an anhydritic Triassic claystone. *Rock Mech. Rock Engng.*, 51(9), 2881–2894.

Ramon, A., Alonso, E., Olivella, S. (2017). Hydro-chemo-mechanical modelling of tunnels in sulfated rocks. *Géotechnique*, 67(11), 968–982.

Sagaseta, C. (1987). Analysis of undrained soil deformation due to ground loss. *Géotechnique*, 37, 301–320.

Sass, I. and Burbaum, U. (2010). Damage to the historic town of Staufen (Germany) caused by geothermal drillings through anhydrite-bearing formations. *Acta Carsologica*, 39(2), 233–245.

Scherer, G.W. (1999). Crystallization in pores. *Cement and Concrete Research*, 29, 1347–1358.

Sherwood, P.T. (1962). Effect of sulphates on cement and lime stabilized soils. *Highwaay Res. Bull.*, 353, 98–107.

Snedker, E.A. (1996). M40: Lime stabilization experiences. In *Lime Stabilization*, Rogers, C.D.F., Glendinning, S., Dixon, N. (eds). Thomas Telford, London.

Steiner, W. (1993). Swelling rock in tunnels: Rock characterization, effect of horizontal stresses and construction procedures. *Int. J. Rock Mech. Min. Sci.*, 30(4), 361–380.

Steiner, W. (2020). Sulphate-bearing rocks in tunnels – Lessons from field observations and in-situ swelling pressures. *Geomechanics and Tunnelling*, 13(3), 286–301.

Steiner, W. and Schwalt, M. (2019). In situ swelling pressure in sulphate bearing rocks. Findings from field observations. In *Proceedings of ISRM 2020: Rock Mechanics for Natural Resources and Infrastructure Development*, Fontoura, R. and Pavón, M. (eds). 1724–1731. ISBN 978-0-367-42284-4.

Wittke, M. (2006). Design, construction, supervision and long-term behaviour of tunnels in swelling rocks. In *EUROCK 2006: Multiphysics Coupling and Long Term Behaviour in Rock Mechanics*, van Cotthem, A., Charlier, R., Thimus, J.-F., Tshibangu, J.-P. (eds). CRC Press, Boca Raton.

Wittke, W. (2014). *Rock Mechanics Based on an Anisotropic Jointed Rock Model (AJRM)*. Ernst, Sohn, Berlin.

Wittke, W., Wittke, M., Wittke-Gattermann, P., Stoffgesetz, E.C. (2017). Berechnungsverfahren, felsmechanische Kennwerte und usführungsstatik für Tunnel im anhydritführenden Gebirge. In *Proceedings of Vortrag anlässlich des 3.* Felsmechanik- und Tunnelbautages im WBICenter, WBI-Print 20, Weinheim.

Wittke-Gattermann, P. and Wittke, M. (2004). Computation of strains and pressures for tunnels in swelling rocks. *Tunnelling and Underground Space Technology*, 19, 422–423.

7

Salts in Heritage Sites

Julie DESARNAUD[1,2]

[1] *Labo des monuments et décoration monumentale, KIK-IRPA, Brussels, Belgium*

[2] *CSTC, Brussels, Belgium*

7.1. Introduction

The alteration of materials under the action of soluble salts (hereinafter referred to as "salts") is a phenomenon known since antiquity. Herodotus, speaking of the Nile Valley, said "salts exude from the ground in such quantities that they even affect the pyramids" (translation of Selincourt in 1972). However, it was not until the 19th century that this type of weathering became a subject of scientific research. These studies lead to a series of hypotheses and models as to the probable deterioration mechanisms involved. This type of deterioration is often observed on traditional building materials (stone, brick, mortar, plaster, etc.), which make up many historical monuments and archaeological sites, with a large quantity of salts accumulating over the centuries. To preserve our cultural heritage, it is therefore necessary to find ways to limit the deterioration of these materials by salts.

This weathering is not specific to stone buildings. Indeed, salts also seem to play a role in the development of geomorphological phenomena such as alveolarization (honeycomb weathering) (Mustoe 1982) and tafoni (Bradley et al. 1978), known under the term haloclasty. They contribute to the production of rock debris, especially in deserts (Goudies 1997) and coastal regions (Mottershead 1982).

For several decades, many studies have shown that weathering is linked to repeating dissolution/recrystallization cycles of salts in porous materials. These phase transitions are strongly influenced by environmental conditions. Due to

Salt Crystallization in Porous Media,
coordinated by Hannelore DERLUYN and Marc PRAT. © ISTE Ltd 2024.

climate change, the intensification of variations in extreme climatic conditions (temperature, relative humidity [RH], rain and drought) is likely to accentuate the degradation of built heritage by salts (Shahidzadeh-Bonn et al. 2010; Espinoza-Marzal and Scherer 2013; Flatt et al. 2014; Desarnaud et al. 2015).

Salts of various origins go through cycles of crystallization/dissolution that have detrimental effects on the materials and take different forms. The different types of alteration caused by salts are listed in the ICOMOS reference book entitled *Illustrated glossary on stone deterioration patterns* (Vergès-Belmin et al. 2008).

There are several methods available to protect our built heritage, which represents the memory of our history and building traditions, from deterioration caused by the crystallization of salts. Some will limit the crystallization of salts in the material and thus reduce degradation without extracting the salts. Others address the problem by desalinating contaminated materials. These are then known as desalination methods by either compression, water bath or electrophoresis.

This chapter describes the alteration of built heritage materials by salts: from the origin of the salts to the preservation of contaminated monuments.

7.2. Salt–Water interaction

Salts are ionic compounds made up of cations (positive ions) and anions (negative ions) and sometimes water molecules in its lattice. These ions can be either mineral (Na^+, Mg^{2+}, etc.), organic ($CH_3\,COO^-$), monoatomic (K^+, Cl^-, etc.) or polyatomic (SO_4^{2-}, NO_3^-). In the solid state, they appear as crystals with ordered and periodic arrangements of atoms and are delimited by a set of faces. For example, halite is a sodium chloride crystal that crystallizes with a face-centered cubic crystalline system. In the liquid state, in water, the ions dissociate, the crystals dissolve to form a saline solution.

7.2.1. *Crystallization*

Salt crystals will precipitate when the concentration of the salt-solution, composed of a mixture of ions, exceeds saturation, which corresponds to the maximum quantity of salt that can be dissolved in a certain volume of water (e.g. 360g/L in water at 20°C for NaCl). A salt-solution is a binary system consisting of a solute (substance that can crystallize) dissolved in a solvent (substance in which the solute is soluble). Crystal growth in solution is an example of a phase transition. A stable phase, the crystal, grows at the expense of an unstable (disordered) phase, the solution. The driving force of the transition is the chemical potential difference $\Delta\mu$ between the two phases:

$$\Delta\mu = [\left(\frac{d\Delta G}{dn}\right)]_{T,P}$$

where ΔG is the change in the free enthalpy of transformation and n is the number of solute molecules. When $\Delta\mu = 0$, there is equilibrium between the two phases of infinite size. When $\Delta\mu > 0$, the phase with the lower chemical potential grows at the expense of the other, until the chemical potentials reach equilibrium ($\Delta\mu = 0$, condition of chemical equilibrium between the two phases). Supersaturation is the driving force of growth. It represents a deviation from equilibrium such that the salt concentration in the solution is greater than its solubility.

Supersaturation of a solution can be achieved in several ways, the most common of which for heritage buildings is by evaporation of the solvent (water). In this case, when the concentration of the solution increases, the chemical potential of the solution at concentration C (T_0) increases with respect to the chemical potential of the initial solution at concentration C_{eq} (T_0). The driving force of crystallization is therefore written as:

$$\Delta\mu - RT_0 . ln\left(\frac{C}{C_{eq}}\right)$$

R being the ideal gas constant and T_0 the temperature in Kelvin.

The degree of supersaturation is also influenced by temperature, by the dynamics of solution transport and evaporation inside a porous material. Consequently, the properties of the salt-solution, in particular its viscosity and its surface tension, have a significant impact (Sghaier et al. 2006; Ruiz-Agudo et al. 2007; Shahidzadeh et al. 2008).

Once the salts have crystallized, if evaporation of the solution maintains a sufficient concentration, the salt crystals formed will continue to grow. The degree of supersaturation reached during crystallization governs the morphology of crystals (Figure 7.1). The higher the degree of supersaturation, the faster the salt crystals will form, and their shape will be much different from their equilibrium shape. Under these conditions, crystals known as *hopper crystal* can form because the concentration of the solution is very high at the corners of the cubic crystal causing growth at the edges before the crystal faces have completely grown (Desarnaud et al. 2018).

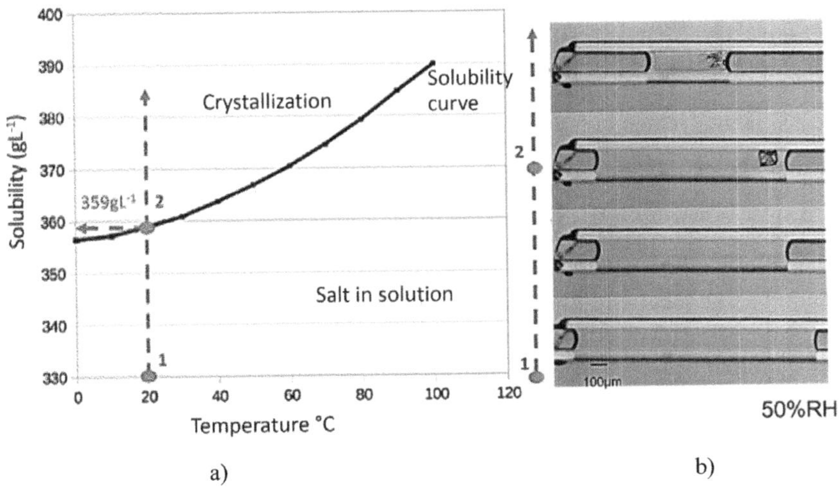

Figure 7.1. *(a) Solubility diagram of NaCl. (b) An undersaturated solution of NaCl which evaporates at 50% RH until the solution exceeds its solubility limit and a NaCl crystal precipitates and grows. For a color version of this figure, see www.iste.co.uk/derluyn/salt.zip*

7.2.2. Dissolution

Salt crystals can dissolve in a solvent. During the introduction of a chemical species initially in the solid state (the solute here in the form of crystals of salt), in a solvent (here liquid, water), the ions of the chemical species dissociate in solution and interact with solvent (water) molecules. This interaction is called solvation. It is of a different nature depending on the solute. A new homogeneous phase is then formed, an electrolytic salt-solution which will, depending on its ion concentration, modify the water–liquid/water–vapor balance. In a closed system containing a volume of water, the liquid phase and the gaseous phase reach an equilibrium, which is a function of the temperature and corresponds to a saturated vapor pressure. The ratio between the partial vapor pressure in the air and the saturated vapor pressure (maximum quantity of water vapor that the air can contain) defines the RH. At a given temperature, in a closed volume containing a volume of salt-solution, the saturation vapor pressure is lower than when it is a volume of pure water. For the same temperature, the more concentrated the salt-solution, the more the saturated vapor pressure decreases to a minimum value when the concentration of the saline solution has reached saturation. This concentration corresponds to the solubility of the salt which is specific to each type of salt (e.g. 360 g/L for sodium chloride and 2.14 g/L for calcium sulfate). This saturated vapor pressure at equilibrium with a saturated saline solution is therefore specific to each salt and defines the relative humidity equilibrium of the salt (RH_{eq}) at a given temperature (Greenspan 1973). Table 7.1 presents the RHeq of the main salts found in heritage buildings. When the humidity of the air is higher than the RHeq of the salt, $RH_{air} > RH_{eq}$, the solution becomes more and more diluted by incorporating water molecules from the air. On the other hand, when $RH_{air} < RH_{eq}$ the salt-solution will concentrate while evaporating until it reaches saturation and exceed the solubility of the salt resulting in crystallization. Thus, variations in the RH conditions of the air will lead to cycles of dissolution/crystallization of the salts, generating alterations.

Type of salt	Chemical formula	Name	Relative humidity of air (%) in equilibrium with the solution saturated in salt at 20°C
Sulfates	Na_2SO_4	Sodium sulfate (thenardite)	82([1])
	$Na_2SO_4.10H_2O$	Sodium sulfate (mirabilite)	94([1])
	$MgSO_4.7H_2O$	Magnesium sulfate (epsomite)	90([2])
	K_2SO_4	Potassium sulfate	97([3])
	$CaSO_4.2H_2O$	Calcium sulfate (gypsum)	> 99([2])
Nitrates	$Mg(NO_3)_2.6H_2O$	Magnesium nitrate (nitromagnesite)	54([3])
	$Ca(NO_3)_2.4H_2O$	Calcium nitrate (nitrocalcite)	54([3])
	NH_4NO_3	Ammonium nitrate	65([3])
	$NaNO_3$	Sodium nitrate (nitratite)	75([3])
	KNO_3	Potassium nitrate (niter)	93([3])
Chlorides	$CaCl_2.6H_2O$	Calcium chloride (antarcticite)	31([3])
	NaCl	Sodium chloride (halite)	75([3])
	KCl	Potassium chloride	85([3])

Table 7.1. *RH equilibrium of the most common salts found in built heritage (extract from NIT 252 of CSTC; Centre scientifique et technique de la construction 2014)*

7.2.3. *Hygroscopicity and deliquescence*

Soluble salts can be hygroscopic, they have the capacity to absorb the water molecules from the air when the RH of the air is higher than the RH equilibrium of salts. If the quantity of moisture adsorbed by the salts is large enough, they will be able to pass into solution: this is the phenomenon of deliquescence. Some salts are slightly hygroscopic such as calcium sulfate (gypsum) (Charola 2007). The lower the relative humidity equilibrium, the greater the hygroscopicity of a salt. When a building is contaminated by a hygroscopic salts and the RH of the air is higher than the salt RH equilibrium damp patches will be visible on the façade (Figure 7.2).

Figure 7.2. *Moisture stains caused by the presence of hygroscopic salts in walls (a) @B. Lubelli; (b) @KIK-IRPA. For a color version of this figure, see www.iste.co.uk/derluyn/salt.zip*

Hygroscopicity, the ability to dissolve, deliquescence and crystallization are important properties in the mechanism of degradation of materials by salts.

7.2.4. *Salt–water–material interaction*

The salts are transported inside the material dissolved in the water through the porous media. Thus, the porosity – here we mean the pore size distribution and the connectivity – will strongly influence the location of salt crystallization as well as the quantity of salts in the material. When the material is not saturated with water and in contact with a source of liquid water, the dissolved salts can penetrate by capillarity. The movement of salts is therefore associated with that of water. Capillarity induced water movement occurs due to capillary pressure (P_c) which is the pressure difference between water (P_l) and air (P_0) separated in the pore by a meniscus:

$$P_C = P_0 - P_l$$

The surface energy of dry pores (γS_0) is greater than the surface energy of the water–saturated pores (γS_l). Since the system decreases in energy when the dry surface is replaced by the wet surface, the capillary forces cause the water to move toward the zones where its free energy will be the lowest (from the wet zones to the dry zones). The maximum height of capillary rise is defined by Jurin's law:

$$H = 2\gamma cos\theta / \rho r g$$

with:

- H: liquid height (m);
- γ: liquid–air surface tension ($N \cdot m^{-1}$);
- θ: contact angle at equilibrium between the liquid and solid (°);
- r: pore radius (m);
- ρ: density of the liquid ($kg \cdot m^{-3}$);
- g: gravitational acceleration ($m \cdot s^{-2}$).

The greater the radius of the pores, the greater the height of the capillary rise. Thus, theoretically in a material with pores of 0.1 µm in diameter, the maximum height of capillary rise is 15 m. In buildings, this height is never reached because it is a balance between the rate of water absorption by capillarity and the rate of water evaporation. If surface coatings do not modify the drying properties of the walls (hydrophobic coatings), capillary fringe heights of around 1.5 m are generally observed. It should be noted that in the case of Saint Mark's Basilica in Venice, very high water contents were measured at more than 6 m in height (Falchia et al. 2018). Liquid water can also enter a porous material by infiltration, the main force involved will then be gravity.

Salts are thus transported dissolved in the water by two mechanisms: advection and diffusion. Advection is the transport of ions by liquid water under the effect of a temperature and/or water content gradient. The second mechanism is diffusion, which corresponds to the transport of ions within liquid water under the effect of a concentration gradient. The diffusion mechanism aims to equalize the concentration of ions at any point in the salt-solution.

In materials with large, well-connected pores, the evaporation of the solution will be greatest, and the salts will crystallize there first. We note that in old masonry, mortars composed mainly of airl lime will degrade first. This is caused firstly, by their low mechanical resistance compared to bricks and/or natural stones, and secondly by their high, well-connected porosity, which facilitates the transport and evaporation of the saline solution, making it a preferential site for the crystallization of salts. It is important to note that when these mortars are restored using mortar based on hydraulic lime or cement, their porosity (pore size distribution, connectivity) is strongly reduced, consequently, the preferential site of salt crystallization becomes stone and/or brick, causing severe damage to buildings. We can thus conclude that the pore size distribution and the connectivity have a significant impact on the location of the salts crystallization and consequently on the degradation.

7.3. Salts in heritage sites

7.3.1. *Nature and origin of salts*

Most of the walls of heritage buildings are contaminated by soluble salts in varying degrees. They are generally concentrated at the surface or subsurface. The main salts are shown in Figure 7.3. These are therefore chlorides, nitrates, carbonates, sulfates, oxalates and more complex salts such as thaumasite. The nature of the salts in masonry sometimes gives indications about their origin. They can come from atmospheric pollution, result from capillary transport from the ground, from the material itself, and from previous unsuitable restoration interventions.

The presence of salts may be inherent in the material. It has been shown that certain quarry stones, such as molasses from eastern Switzerland, can contain high levels of soluble salts (sulfate in the case of molasses; Charola and Blaüer 2015). Some lime mortars made from dolomite limestone naturally contain magnesium limestone. After percolation of water through the mortar, the magnesium limestone can react with sulfate ions and form magnesium sulfate (epsomite $MgSO_4 \cdot 7H_2O$). Portland cements may contain a significant amount of sodium oxides which, after reacting with sulfuric acid from polluted air, can form sodium sulfate (thenardite: Na_2SO_4 and mirabilite: $Na_2SO_4 \cdot 10H_2O$) (Bläuer-Böhm and Häfner 1996). In contact with certain calcareous stones, sodium oxides can form carbonates such as natrite ($NaCO_3^- \cdot 10H_2O$).

Sea brine is one of the sources of chloride ions and sodium ions in buildings (Lubelli et al. 2001; Klenz 2007). This contamination is not limited to coastal regions, indeed NaCl from sea salts has been found up to 900 km inland. A large part of the salts comes from human activities. De-icing salts (mainly antarcticite: $CaCl_2 \cdot 6H_2O$, and halite: NaCl) spread on the ground in winter, once dissolved in water will strongly contaminate buildings. The water that rises by capillarity in masonry will carry with it the salts contained in the soil. Thus, in rural areas or close to cemeteries, masonry is contaminated by nitrates which very often come from the oxidation of organic matter (Vergès-Belmin 1992). Close to agricultural areas, buildings are frequently loaded with potassium (sylvite KCl) and sodium nitrate (nitratine $NaNO_3$) originating from pesticides and fertilizers, respectively.

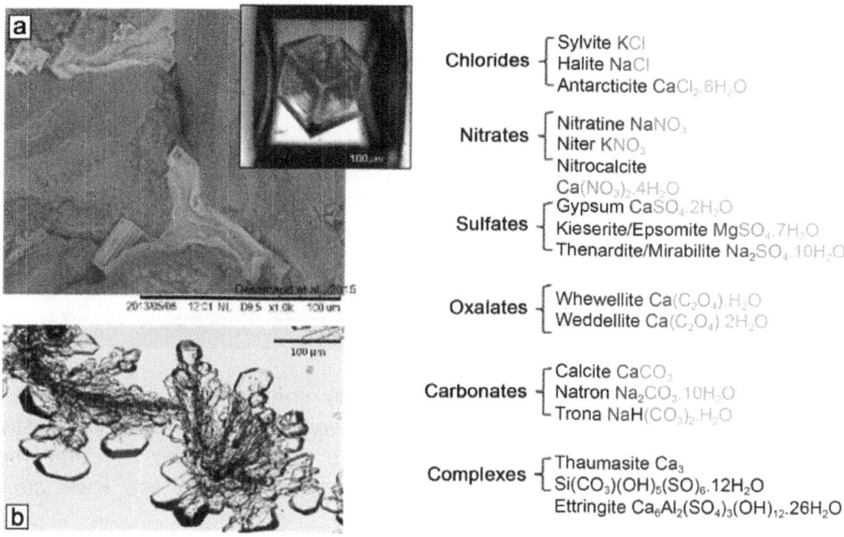

Figure 7.3. *Main salts present in buildings. (a) Image obtained by scanning electron microscopy of NaCl crystals in the intergranular space of a Prague sandstone (Desarnaud et al. 2015). Inset details of a cubic crystal of NaCl grown in a glass capillary tube. (b) Crystallization of thenardite (dendrites) and mirabilite (Desarnaud and Shahidzadeh 2012). For a color version of this figure, see www.iste.co.uk/derluyn/salt.zip*

Another significant source is the products used for the conservation and restoration of building stones. Many of them commonly used accelerate the degradation of building materials by salts weathering. Many alkaline silicate based products with hydrophobic properties have long been used for wall injections and as consolidation products. The use of sodium and potassium silicate as consolidation treatments tends to form very hard layers that come off over time. After reacting, alkaline silicates easily form sodium and potassium carbonates, hygroscopic salts, which tend to form damp patches on facades. Sulfates in polluted air tend to form sodium sulfates (thenardite: Na_2SO_4 and mirabilite: $Na_2SO_4 \cdot 10H_2O$) and potassium sulfates (arcanite: K_2SO_4), which decay porous materials. Cleaning stones with NaOH has often resulted in the formation of alkaline carbonates (Bromblet 2004).

Some living organisms can, given their metabolism, naturally produce salts. Sulfates can be caused by the presence of sulfur-oxidizing bacteria (example of the Bethlehem altarpiece in the cathedral of Narbonne; Rager et al. 1996).

Figure 7.4. *(a) Black crusts of gypsum formed by condensation on marble in an urban environment (bust of Poelaert, Laeken cemetery, Belgium). (b) Black crusts of gypsum on limestone causing degradation of the support during its detachment (Laeken cemetery, Belgium). For a color version of this figure, see www.iste.co.uk/derluyn/salt.zip*

Finally, air pollution in the urban environment will intensify the presence of sulfates, sulfuric acid and sulfur dioxide, which mainly manifest themselves by forming gypsum crusts ($CaSO_4 \cdot 2H_2O$) (Steiger et al. 2014). They are usually black in color because they encapsulate pollutant particles (such as soot for example). These crusts are visible in sheltered areas of a building (Figure 7.4).

7.3.2. *Rise and distribution of salts in masonry*

Salt solution will percolate in the masonry and will to crystallize where it evaporates. The least soluble salts will thus crystallize first, while the most soluble salts will crystallize last. During capillary rise of the salt solution in the walls, the less soluble salts will crystallize at the base of the wall, while the more soluble salts will precipitate on higher parts of the wall. The phenomena observed in this situation are shown in Figure 7.5 (Zehnder 1989). The wall is subject to rising damp from the ground containing the most commonly encountered ions (the cations: sodium (Na^+), potassium (K^+), magnesium (Mg^+), calcium (Ca^+); the anions: chloride (Cl^-), sulfate (SO_4^{2-}), nitrate (NO_3^-), bicarbonate ($HCO3^-$)). Less soluble salts, such as calcium carbonate ($CaCO_3$) and gypsum ($CaSO_4.2H_2O$), will form first near the ground (zone A). The salt-solution containing the other ions continues to rise in the wall by capillarity while evaporating. Thus, other salts crystallize such as epsomite ($MgSO_4·7H_2O$) or hexahydrite ($MgSO_4·6H_2O$) depending on weather conditions. Thenardite or mirabilite (Na_2SO_4 and $Na_2SO_4·10H_2O$) crystallizes depending on the temperature. Niter (KNO_3) will crystallize at low temperatures because its solubility is strongly temperature dependent.

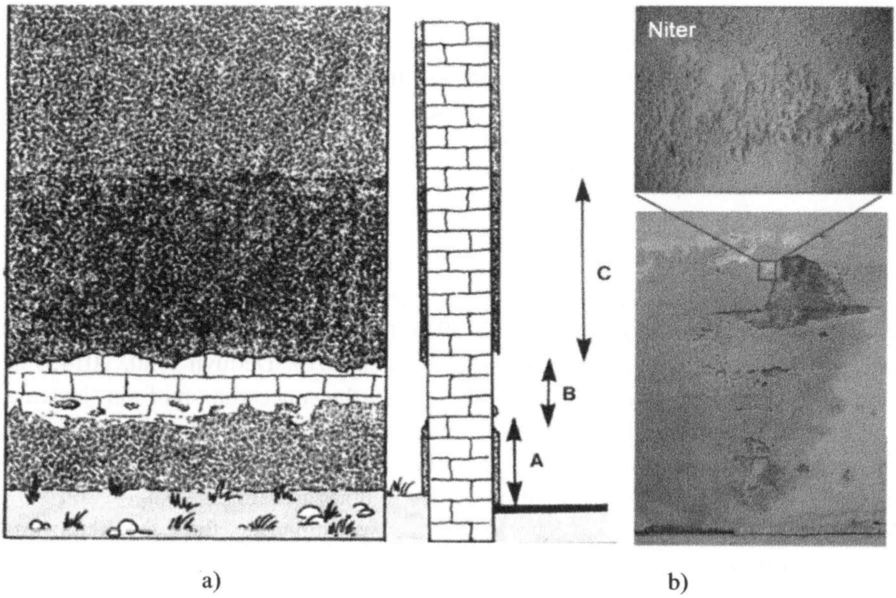

Figure 7.5. *(a) Illustration of the distribution of salts in a wall subject to to capillary rise (Zehnder 1979). (b) Photograph of niter efflorescence and dark patches caused by hygroscopic salts (Mertz 2007). For a color version of this figure, see www.iste.co.uk/derluyn/salt.zip*

Zone C generally appears darker because it is more humid with little efflorescence (unless the atmosphere is very dry or indoors). It is at this level that the most hygroscopic salts are concentrated, such as nitronatrite ($NaNO_3$) and niter (KNO_3), halite (NaCl) and sylvite (KCl). Most of the time these salts remain in solution because they retain the humidity contained in the wall or absorb the water molecules from the air. Thus, they will slow down evaporation and modify the transport of moisture. It is important to know that even after treatment against rising damp and drying of the walls, these stains may persist due to the hygroscopic properties of these salts. In zone C, the salts will crystallize when the humidity of the ambient air is low, less than 63% for sodium chloride (Godts et al. 2021). This value is well below 75%, the RH equilibrium of halite, because when several salts are in solution, their solubility change. For less soluble salts, such as gypsum or calcium carbonate, the presence of other salts, such as calcium and potassium chloride, will increase their solubility (Leithner et al. 2005). The salts in zone C will, therefore, crystallize and dissolve regularly, either seasonally or daily depending on climatic changes. This frequency of dissolution/crystallization cycles is the cause of the deterioration of materials often visible on a facade in areas B and C.

The precipitation order of salts in a mixture has been studied in detail by geologists in the context of evaporation of water from lakes (formation of evaporites) (Goudies and Viles 1997). Models to predict crystallization sequences have also been developed (Steiger 2005).

7.3.3. *Weathering mechanisms*

The salts will crystallize either on the surface of the materials and form salt efflorescence (Veran-Tissoire et al. 2012, 2014) (Figure 7.6) either beneath the surface and form subflorescence (Shahidzadeh-Bonn et al. 2010; Derluyn et al. 2014; Flatt et al. 2014). As the water evaporates at the surface, the concentration of ions increases until it exceeds the solubility of the salt. There will then be nucleation, crystallization and growth of salt crystals at the evaporation front. The crystallization site on the substrate will depend on the migration of the solution toward the evaporation zone. When the water flux is slower than the water evaporation rate, the evaporation front moves inward from the surface and subflorescence is formed. Thus, the environmental conditions modify the transport properties of the saline solution and the site of salt crystallization. High-humidity conditions have been shown to favor the formation of efflorescence, while low-humidity conditions favor the formation of subflorescence (Rodriguez-Navarro 1999). Also note that the salt concentration of the solution will have an impact on its viscosity, its density, its surface tension and consequently influence its transport in the porous medium (Sghaier et al. 2006; Shahidzadeh et al. 2008). These conditions are discussed in Chapter 3 of this book.

Efflorescence causes only minor damage, as they simply form on the surface and can be easily removed by dry brushing. The morphology of the crystals (rhombohedral, acicular crystals, etc.) in the efflorescence provides information on their crystallization conditions (Zehnder 1989). Once formed, the efflorescence can adhere strongly, limiting their removal (this is the case of NaCl) (Desarnaud et al. 2015). Depending on the type of crystallization, the efflorescence can modify the evaporation properties of the saline solution. If they form a crust on the surface, they will slow down the evaporation flux; on the other hand, if they form cauliflower-like crystals, they will increase this flux (Figure 7.7) (Sghaier and Prat 2009; Eloukabi et al. 2013; Veran-Tissoire et al. 2014; Desarnaud and Shahidzadeh 2015).

Figure 7.6. *(a) White salt efflorescence on the surface of sandstone, Figeac cathedral, France (de-icing salt) (Bromblet 2010). (b) Efflorescence of NaCl in the form of "wiskers", which corresponds to slow growth in conditions of high humidity, Ostend water tower, Belgium (sea salt). For a color version of this figure, see www.iste.co.uk/derluyn/salt.zip*

Figure 7.7. *(a) Formation of thick black crusts of gypsum. (b) Crusts formed by the transition from dry to wet deposition on limestone (Maison du Roi, Brussels, Belgium). For a color version of this figure, see www.iste.co.uk/derluyn/salt.zip*

Figure 7.8. *(a) Granular disaggregation (Belgium). (b) Honeycomb weathering on sandstone by the sea (Spain). For a color version of this figure, see www.iste.co.uk/derluyn/salt.zip*

The crystallization of salts in porous material under the surface will cause damage to the materials. The different types of alterations generated are listed in the ICOMOS illustrated glossary (Vergès-Belmin 2008). The most common alterations are granular disintegration, which corresponds to the crystallization of salts in the intergranular space of the stone causing the grains to separate (Figure 7.8(a)). This type of alteration can be superficial or deep and is encountered on masonry made up of granular stone (such as sandstone) or with a grainy texture (such as granite). Depending on the characteristics of the stone, this type of alteration can lead to damage such as honeycomb weathering (Figure 7.8(b)). When crystallization is deeper, the deterioration will appear as sub-millimeter to millimeter sized flakes (Figure 7.9) or contour scaling. In this case, layers parallel to the surface will form and eventually detach. This is the case in Figure 7.9(b) where the cherub head has lost most of these sculpted details due to crystallization below the surface.

Many articles found in the literature highlight the strong interest in understanding the alteration of stone by salt weathering (Desarnaud et al. 2013, 2016). Many hypotheses have been put forward as to the probable mechanisms involved. Among these hypotheses, we find the difference in thermal and hygric expansion between the salt and the porous material, the hydration pressure that would develop in a porous medium with the increase in volume of the salt crystal during its hydration. Mortensen (1933), Rodriguez-Navarro and Doehne (1999),

Flatt (2002) and Scherer (1999, 2000) have shown that the hydration pressure does not exist as such and that it is just a special case of the crystallization pressure. This last hypothesis, the crystallization pressure caused by a salt crystal growing in a confinement, has been extensively studied in recent years (Hamilton et al. 2010; Sekine et al. 2011; Royne and Dysthe 2012; Desarnaud et al. 2013, 2016; Naillon et al. 2018; Dziadkowiec et al. 2019) and is the topic of Chapter 2 of this book. The pressure generated could reach values higher than the mechanical resistance of stones and explain their degradation.

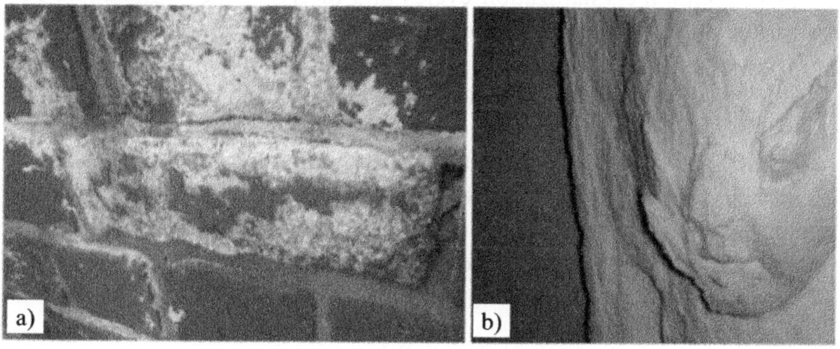

Figure 7.9. *(a) Crystallization of salt on the surface of the brick after repairing old mortars with cement-based mortars, leading to the formation of scaling. (b) Detachment of thin layers of marble parallel to each other and parallel to the surface. For a color version of this figure, see www.iste.co.uk/derluyn/salt.zip*

7.4. Treatment of objects/monuments contaminated by salts

Before being able to treat a highly degraded sculpture or building, it is necessary to make a diagnosis in order to determine the origin, the type and the amount of salts present as well as the kinetics of degradation. This involves collecting information from the people in charge of the building, visual observations and measurements. Measurements will be carried out on samples taken either from the surface of the material (efflorescence) or from within the material by drilling. It is important to note here that depending on the external climatic conditions, the results may vary. It is therefore necessary to compare these results with the temperature, humidity and precipitation on the day of sampling, especially if these results are used as support for a comparison before and after treatment. After extraction (NF EN16455:2014), the ions are quantified and identified. Several field-based and laboratory-based techniques exist. On site, the use of strips that change color on contact with certain ions is a simple approach. Laboratory analyses make it possible to more accurately identify the efflorescent salts (X-ray diffraction) and to identify and quantify the ions in the powders taken from the sample (SEM-EDX, ion chromatography). Ion concentration values, above which degradation is caused by the crystallization of salts, have been published in the literature (Verges-Belmin 1992). These values are indicative as they do not consider other important factors such as those related to the stone: pore size distribution, porosity and pore connectivity. The salt concentrations are also not compared with the climatic conditions; many variations increase the number of dissolution/crystallization cycles and therefore the degradation. Finally, the interaction between the ions is not considered, although it has been shown that certain mixtures of salts have a much more destructive effect than each salt taken individually (Robinson and William 2000; William and Robinson 2001).

The results of the analyses will often make it possible to identify the salts, their origin and their degradation kinetics. If the salt crystallizations are likely to aggravate the degradation of the materials, two treatments are possible: (i) stabilization of the environmental conditions; (ii) the extraction of salts from the object.

7.4.1. *Stabilizing environmental conditions*

The results of the contents and the identification of the ions present obtained do not provide any information about the behaviour of such salt-solutions as a function of climatic conditions. Which salts will precipitate? At what humidity and at what temperature will they precipitate? The use of a thermodynamic model (Steiger and Zeunert 1996; Price 2000), which describes the behavior of salt mixtures at equilibrium, makes it possible to predict under which temperature and humidity conditions certain salts will crystallize. If the environmental conditions are stable, this makes it possible to prevent the deterioration of materials by salt crystallization. This was the case during the study of murals in the Mespelaere church in Belgium (Crevals et al. 2021). These paintings dating from the 14th to the 19th century were affected by salt crystallization and rising damp, as well as the masonry below. Salt efflorescence affect the reading of the wall paintings and crystallization in the form of subflorescence caused the peeling of the paint and the granular disintegration of the masonry brickwork (Figures 7.10(a) and (b)). The analyses revealed a very large quantity of salt, essentially NaCl and Na_2SO_4, impossible to extract using desalination poultices. The analysis of the behavior of the ionic mixture measured in the brick, the mortar and the plaster of the mural supports thanks to the program

ECOS RUNSALT (Price 2000; Bionda 2005) showed that most salts crystallized between 60% and 80% RH, and that no salt crystallized between 45% and 60% RH; below 45% RH, only a tiny amount of salt crystallized (Figure 7.10(c)).

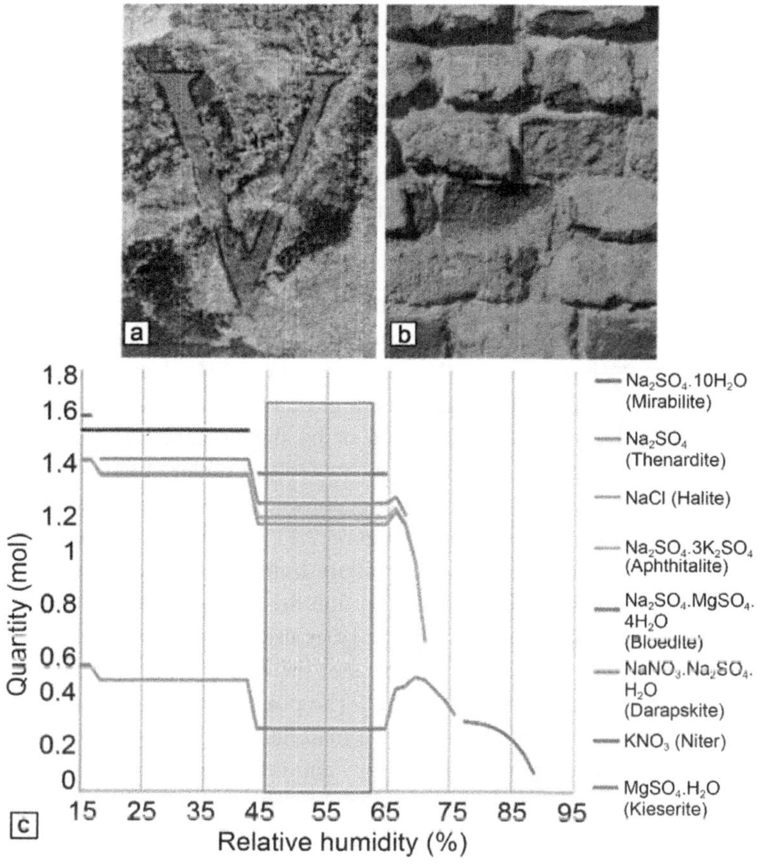

Figure 7.10. *(a) A section of 16th century wall painting partially covered by salt efflorescence. (b) Granular disintegration of the masonry under the murals. (c) Hydrothermal behavior of the ion mixtures measured in the mortar. The box in blue shows that, in the zone with 45% to 60% relative humidity, there is no crystallization of salts (Crevals et al. 2021). For a color version of this figure, see www.iste.co.uk/derluyn/salt.zip*

In this case, once the source of the salts had been eliminated by a treatment to prevent rising damp, the environmental conditions inside the church were stabilised at a humidity level of less than 60% to prevent the deterioration of the wall paintings. Maintaining the humidity below 60% stops most of the dissolution/crystallization of the mixture of salts present, and therefore the degradation of the paintings.

7.4.2. *Salt extraction*

The extraction can be done by the application of a wet poultice against an object or masonry, desalination by baths can be used to treat objects and the application of electrophoresis can be done on masonry.

Desalination poultices, made of one or more hydrophilic materials (cellulose, clay, etc.), are applied to the surface of masonry and have been the focus of numerous studies (Vincente and Vincente-Tavera 2001; Vergès-Belmin 2003; Vergès-Belmin and Siedel 2005). The salts migrate from the masonry to the poultice either by diffusion of ions or by advection or by their combined action. Diffusion occurs when ions migrate through wet materials under the influence of a concentration gradient from the highest concentration (in the masonry) toward the lowest concentration (in the poultice) (Figure 7.11). Advection occurs when ions migrate through the porous material by capillary transport along with the solution in which they are found (Figure 7.11). It is through this phenomenon of advection that desalination occurs in the poultice when the latter dries by evaporation. For this type of desalination to work, the size of the pores of the poultice must be adapted to the size of the pores of the salt loaded material. The poultice must have larger pores so that the water migrates by capillarity into the material and dissolves the salts present. The poultice must also contain smaller pores than the salt loaded material so that the salt-solution formed in the material is taken by the poultice. Thus, the ions will migrate by advection into the dryong poultice. If the poultice does not contain pores smaller than the material, the ions risk being redistributed within the masonry. Note here the challenge of desalination of stone with very small pores such as marble by the application of poultices. The poultice must, in addition to the specific pore sizes, be easily applicable to the support and adhere to the support when it dries.

The extraction of salts in movable objects can be done using water baths (Figure 7.12). The object must be able to be moved in order to be partially or totally immersed in tanks filled with demineralized water (less aggressive than distilled water). The material must be able to withstand immersion, which must be verified in prior tests. Once immersed, the water will penetrate the porosity of the material, dissolve the salts and, due to a difference in concentration between the salt-solution newly formed in the object and the water in the bath, the ions will migrate by diffusion toward the water in the bath. This phenomenon is described by Fick's law:

$$J = -D\nabla C$$

where J is the mass flux, D is the diffusion coefficient and ∇C is the concentration gradient. We directly notice the importance of the connection of the pores between them and the tortuosity. It is a process that is very slow because it depends on the penetration of water into the object but also on the difference in concentration between the ionic solution formed after dissolving the salts and the demineralized water. When this difference is large, the diffusion of ions occurs, and it will decrease as the ion concentrations in the object and in the bath equalize. Thus, replenishing the water in the bath improves desalination efficiency but it is costly.

Desalination is usually controlled by measuring the conductivity of the water in the bath (Figure 7.12). The more ions the water contains, the more it conducts electricity. During desalination, the conductivity of the water in bath increases because the ions migrate in the bath water until they reach a threshold, corresponding to the moment when the water in the baths should be replenished. If the conductivity of the water bath is high, the water must be renewed until the conductivity value reached is below the acceptable threshold. This technique is very effective but can be time consuming. The desalination of the Carrara marble blocks of the 400 kg Saint Lazarus altar keystone, sculpted by Francesco Laurana, required 5 baths and 40 months of desalination.

The last method is desalination by electrophoresis. This method is based on the creation of an electrical potential through the masonry. The ions in solution contained in the pores of the materials tend to migrate, depending on their electrical charge, toward the positive electrode or toward the negative electrode, after which they can be collected and removed. These electrodes are placed outside the material; at the interface with the material, there is a buffer made with cellulose which stores the salt ions and will be removed at the end of the desalination.

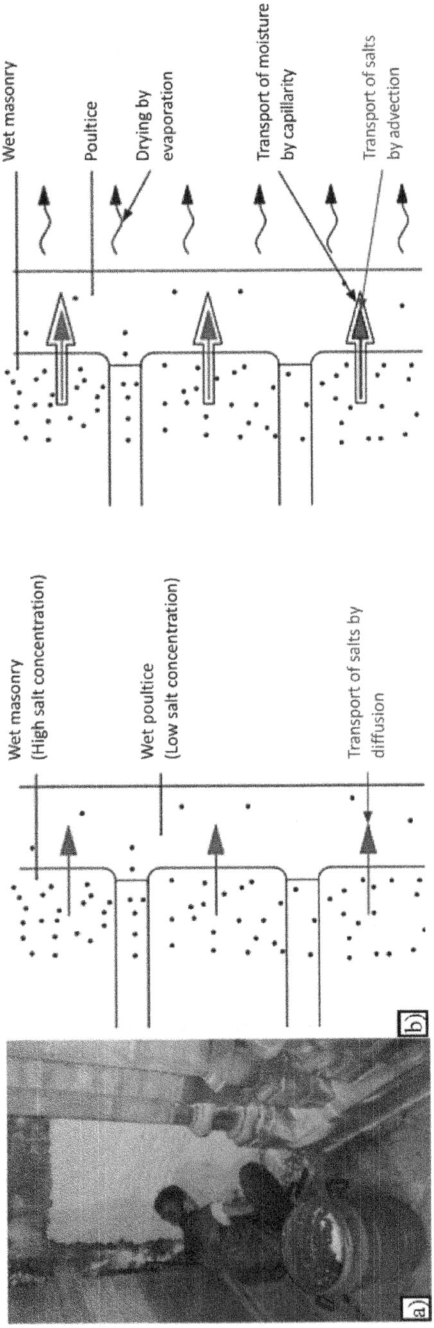

Figure 7.11. *(a) Application of a poultice based on kaolinite cellulose and sand (KIK-IRPA). (b) Mechanisms of salt extraction by application of a hydrophilic poultice on masonry. For a color version of this figure, see www.iste.co.uk/derluyn/salt.zip*

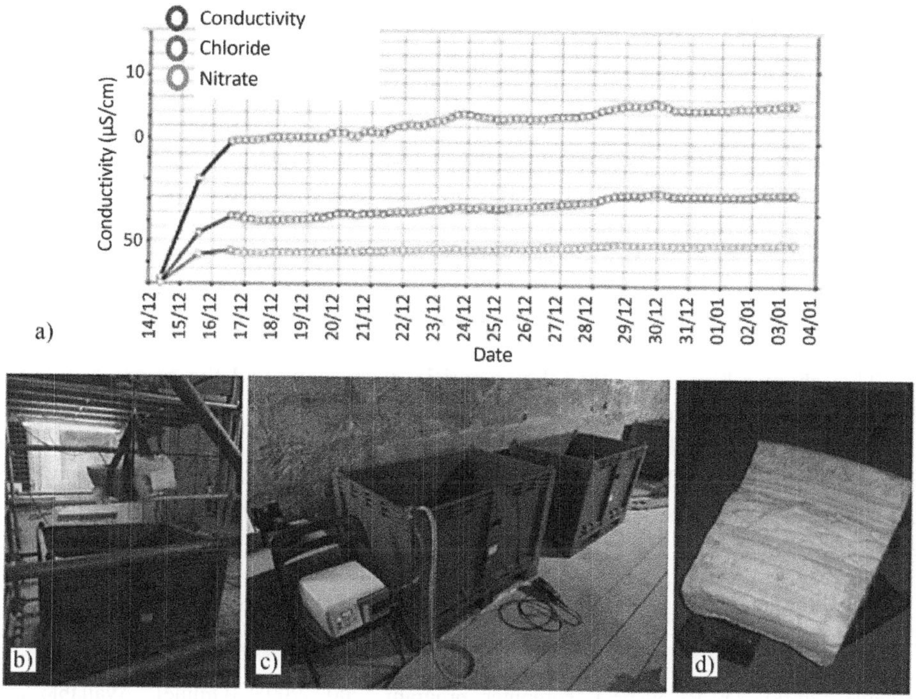

Figure 7.12. *(a) Conductivity curves reaching a threshold, indicating the moment the bath water should be renewed. (b) Bathing of a block of marble from the altarpiece of Saint Lazarus by Francesco Laurana. (c) Desalination by conductivity. (d) Keystone block covered with a poultice for pre-consolidation before bathing (@Sébastien Aze). For a color version of this figure, see www.iste.co.uk/derluyn/salt.zip*

7.5. Conclusion

In this chapter, we have dealt with the alteration of porous materials by salt crystallization. This is one of the main causes of the degradation of monuments. Salts contaminating heritage materials can come from various sources such as air pollution or soil leaching. As they crystallize, they develop pressures that can exceed the mechanical strength of the materials and cause degradation. The types of alteration according to the material, the environmental conditions and the type of salts are listed in the ICOMOS illustrated glossary. After an exhaustive diagnosis has been carried out to identify the salts, their sources, and to measure their effect on the degradation process, interventions must be considered. It is then possible to treat the building either by stabilizing the environmental conditions or by extracting the salts from the material. This last intervention is delicate and can be undertaken by the application of poultices, by immersion in demineralized water baths or using electrophoresis. In all cases, the sources of salts must be stopped before any other intervention. As this chapter shows, the restoration of monuments contaminated by salts is a collective action between restorers, conservators and scientists.

7.6. Acknowledgments

This chapter brings together the scientific discussions and the research carried out by several colleagues. The author thanks Dr. Sebastiaan Godts from the Royal Institute for Cultural Heritage (KIK-IRPA) in Belgium, Pr. Dr. Daniel Bonn and Pr. Dr. Noushine Shahidzadeh from the Institute of Physics of the University of Amsterdam, and Yves Vanhellemont from Buildwise for their contributions and scientific discussions.

7.7. References

Bionda (2005). D. RUNSALT computer program, 2002–2005 [Online]. Available at: http://science.sdf-eu.org/runsalt/.

Bläuer-Böhm, C. and Häfner, K. (1996). Desalting a wall painting after application of silicic acid ester. In *Le dessalement des matériaux poreux : 7es journées d'études de la SFIIC*, Poitiers, May 9–10, 1996. French section of the International institute for conservation of historic and artistic works, Champs-sur-Marne.

Bradley, W.C., Hutton, J.T., Twidale, C.R. (1978). Role of the salts in the development of granitic tafoni, South Australia. *Journal of Geology*, 86, 647–654.

Bromblet, P. (2010). *Guide "altérations de la pierre"*. Association MEDISTONE, Montélimar.

Bromblet, P. and Floc'h, P. (2004). Le retable sculpté de l'église Sainte-Croix de Quimperlé (Finistère) : mise en évidence des interventions antérieures et de leur impact sur les dégradations. In *Retables in situ : conservation et restauration. 11es journées d'études de la SFIIC*, Roubaix, June 24–25, 2004. French section of the International institute for conservation of historic and artistic works, Champs-sur-Marne.

Centre scientifique et technique de la construction (2014). L'humidité dans les constructions. Particularités de l'humidité ascensionnelle. Technical note, 252 [Online]. Available at: https://www.cstc.be/publications/notes-d-information-technique/252/.

Charola, A.E. and Blauer, C. (2015). Salts in masonry: An overview of the problem. *Restoration of Buildings and Monuments*, 21(4–6), 119–135.

Charola, A.E., Pühringer, J., Steiger, M. (2007). Gypsum: A review of its role in the deterioration of building materials. *Environmental Geology*, 52(2), 207–220.

Crevals, V., Godts, S., Desarnaud, J. (2021). Salt problem and climate control in the case of the church of Sint Aldegondis, Mesepelare, Belgium. *5th International Conference on Salt Weathering of Buildings and Stone Sculptures*. Delft.

Derluyn, H., Moonen, P., Carmeliet, J. (2014). Deformation and damage due to drying-induced salt crystallization in porous limestone. *Journal of the Mechanics and Physics of Solids*, 63, 242–255.

Desarnaud, J., Grauby, O., Bromblet, P., Vallet, J.M., Baronnet, A. (2013). Growth and dissolution of crystal under load: New experimental results on KCl. *Crystal Growth & Design*, 13(3), 1067–1074.

Desarnaud, J., Derlyun, H., Molari, L., de Miranda, S., Cnudde, V., Shahidzadeh, N. (2015). Drying of salt contaminated porous media, effect of primary and secondary nucleation. *J. App. Phys.*, 118, 114901.

Desarnaud, J., Bonn, D., Shahidzadeh, N. (2016). The pressure induced by salt crystallization in confinement. *Scientific Reports*, 6, 30856.

Desarnaud, J., Derluyn, H., Carmeliet, J., Bonn, D., Shahidzadeh, N. (2018). Hopper growth of salt crystals. *Journal of Physical Chemistry Letters*, 9(11), 2961–2966.

Dziadkowiec, J., Zareeipolgardani, B., Dysthe, D.K., Røyne, A. (2019). Nucleation in confinement generates long-range repulsion between rough calcite surfaces. *Scientific reports*, 9(1), 1–15.

Eloukabi., H., Sghaier, N.N., Ben Nasrallah, S., Prat, M. (2013). Experimental study of the effect of sodium chloride on drying of porous media: The crusty–patchy efflorescence transition. *International Journal of Heat and Mass Transfer*, 56, 80–93.

Espinosa-Marzal, R. and Scherer, G.W. (2013). Advances in understanding damage by salt crystallization. *Environ. Earth Sci.*, 69, 2657–2669.

Falchia, L., Slanzia, D., Ballianaa, E., Driussib, G., Zendri, E. (2018). Rising damp in historical buildings: A Venetian perspective. *Building and Environment*, 131, 117–127.

Flatt, R.J. (2002). Salt damage in porous materials: How high supersaturations are generated. *Journal of Crystal Growth*, 242(3/4), 435–454.

Flatt, R.J., Caruso, F., Sanchez, A.M.A., Scherer, G.W. (2014). Chemo-mechanics of salt damage in stone. *Nature Communications*, 5(1), 1–5.

Godts, S., Orr, S.A. Desarnaud, J., Steiger, M., Wilhelm, K., De Clerq, H., Cnudde, V., De Kock, T. (2021). NaCl-related weathering of stone: The importance of kinetics and salt mixtures in environmental risk assessment. *Heritage Science*, 9(1), 44.

Goudie, A. and Viles, H. (1997). *Salt Weathering Hazards*. Wiley, New York.

Greenspan, L. (1977). Humidity fixed points of binary saturated aqueous solution. *Journal of Research of the National Bureau of Standards*, 81(1), 89–96.

Hamilton, A., Koutsos, V., Hall, C. (2010). Direct measurement of salt–mineral repulsion using atomic force microscopy. *Chemical Communications*, 46(29), 5235–5237.

Heindrick, S. and Vanhellemont, Y. (2015). L'humidité dans la construction. Technical information note 252, Centre scientifique et technique de la construction, Brussels.

Klenz, L.P. (2007). The salt decay of Medieval bricks at a vault in Brarup church, Denmark. *Environmental Geology*, 52(2), 259–267.

Lubelli, B., van Hees, R.P.J., Groot, C.J.W.P. (2004). The role of sea salts in the occurrence of different damage mechanisms and decay patterns on brick masonry. *Construction and Building Materials*, 18(2), 119–124.

Mertz, J.D. (2007). Altération des matériaux du patrimoine : influence des contaminants et de la nature des substrats. *Journées Scientifiiques de l'environnement (Environnement, Citoyenneté et Territoires Urbains)*, May 3.

Mortensen, H. (1933). Die "Saltzprengung" und ihre Bedeutung für due Regionalklima-tische Gliederung der Wűsten. *Dr. A. Petermanns Mitteilungen*, 79, 130–135.

Mottershead, D.N. (1982). Coastal spray weathering of bedrock in the supratidal zone at East Prawle, South Devon. *Field Studies*, 5, 663–684.

Mustoe, C.E. (1982). The origin of the honeycomb weathering. *Bulletin of the Geological Society of America*, 93, 108–115.

Naillon, A., Joseph, P., Prat, M. (2018). Ion transport and precipitation kinetics as key aspects of stress generation on pore walls induced by salt crystallization. *Physical Review Letters*, 120(3), 034502.

Price, C. (2000). *An Expert Chemical Model for Determining the Environmental Conditions Needed to Prevent Salt Damage in Porous Materials*. Archetype Publications, London.

Rager, G., Payre, M., Lefèvre, L. (1996). Mise au point d'une méthode de dessalement pour des sculptures du XIVe siècle en pierre polychromée. In *Le Dessalement des matériaux poreux*. SFIIC Editions, Paris.

Robinson, D.A. and Williams, R.B.G. (2000). Experimental weathering of sandstone by combinations of salts. *Earth Surface Processes and Landforms*, 25(12), 1309–1315.

Rodríguez-Navarro, C. and Doehne, E. (1999). Salt weathering: Influence of evaporation rate, supersaturation and crystallization pattern. *Earth Surface Processes and Landforms*, 24, 191–209.

Røyne, A. and Dysthe, D.K. (2012). Rim formation on crystal faces growing in confinement. *Journal of Crystal Growth*, 346(1), 89–100.

Ruiz-Agudo, E., Mees, F., Jacobs, P., Rodríguez-Navarro, C. (2007). The role of saline solution properties on porous limestone salt weathering by magnesium and sodium sulfates. *Environmental Geology*, 52(2), 305–317.

Scherer, G.W. (1999). Crystallization in pores. *Cement and Concrete Research*, 29(8), 1347–1358.

Scherer, G.W. (2000). Stressfrom crystallization of salt in pores. In *9th International Congress on Deterioration and Conservation of Stone*, Fassina, V. (ed.). Elsevier, Amsterdam [Online]. Available at: https://doi.org/10.1016/B978-044450517-0/50100-8.

Sekine, K., Okamoto, A., Hayashi, K. (2011). In situ observation of the crystallization pressure induced by halite crystal growth in a microfluidic channel. *American Mineralogist*, 96(7), 1012–1019.

Sghaier, N., Prat, M., Nasrallah, S.B. (2006). On the influence of sodium chloride concentration on equilibrium contact angle. *Chemical Engineering Journal*, 122, 47–53.

Shahidzadeh-Bonn, N. and Desarnaud, J. (2012). Damage in porous media: Role of the kinetics of salt (re)crystallization. *The European Physical Journal Applied Physics*, 60(2), 24205.

Shahidzadeh-Bonn, N., Rafai, S., Bonn, D., Wegdam, G. (2008). Salt crystallization during evaporation: Impact of interfacial properties. *Langmuir*, 24(16), 8599–8605.

Shahidzadeh-Bonn, N., Desarnaud, J., Bertrand, F., Chateau, X., Bonn, D. (2010). Damage in porous media due to salt crystallization. *Physical Review E*, 81(6), 066110.

Steiger, M. (2005). Salts in porous materials: Thermodynamics of phase transitions, modeling and preventive conservation. *Restoration of Buildings and Monuments: An International Journal*, 11(6), 419–431.

Steiger, M. and Zeunert, A. (1996). Crystallization properties of salt mixtures: Comparison of experimental results and model calculations. In *Eight International Congress on the Deterioration and Conservation of Stone*, Riederer, J. (ed.). Möller Druck, Berlin.

Steiger, M., Charola, A.E., Sterflinger, K. (2014). Weathering and deterioration. In *Stone in architecture*, 5th edition, Siegesmund, S. and Snethlage, R. (eds). Springer, Berlin.

Veran-Tissoires, S. and Prat, M. (2014). Evaporation of sodium chloride solution from a saturated porous medium with efflorescence formation. *J. Fluid Mech.*, 749, 701–749.

Veran-Tissoires, S., Marcoux, M., Prat, M. (2012). Discrete salt crystallization at the surface of a porous medium. *Physical Review Letters*, 108(5), 054502.

Vergès-Belmin, V. (1992) . Répartition des sels et cartographie des altérations sur la façade de l'église Notre Dame-La-Grande à Poitiers, France. In *7th International Congress on Deterioration and Conservation of Stone*, Siegesmund, S., Weiss, T., Vollbrecht, A. (eds).

Vergès-Belmin, V. (2006). Desalination of porous building materials: A review. In *Proceedings of the ARCCHIP Workshops, European Research on Cultural Heritage, State-of-the-Art Studies, vol. V*, Simon, S. and Drdacky, M. (eds). Institute for Theoretical and Applied Mechanics, Academy of Sciences of the Czech Republic, Prague.

Vergès-Belmin, V. (ed.) (2008). Illustrated glossary on stone deterioration patterns. *ICOMOS* [Online]. Available at: http://openarchive.icomos.org/id/eprint/434/1/Monuments_and_Sites_15_ISCS_Glossary_Stone.pdf.

Vergès-Belmin, V. and Siedel, H. (2005). Desalination of masonries and monumental sculptures by poulticing: A review. *Restoration of Buildings and Monuments: An International Journal*, 11(6), 391–407.

Vicente, M.A. and Vicente-Tavera, S. (2001). Clay poultices in salt extraction from ornamental stones: A statistical approach. *Clays and Clay Minerals*, 49(3), 227–235.

Williams, R.B.G. and Robinson, D.A. (2001). Experimental frost weathering of sandstone by various combinations of salts. *Earth Surface Processes and Landforms*, 26, 811–818.

Zehnder, K. and Arnold, A. (1989). Crystal growth in salt efflorescence. *Journal of Crystal Growth*, 97(2), 513–521.

8

Salt Crystallization in a Changing Environment

Beatriz MENÉNDEZ
GEC, CY Cergy Paris Université, Neuville-sur-Oise, France

8.1. Introduction

Salts are very widely present in all porous construction materials, both in cities and the countryside. Their chemical composition, degree of hydration and crystalline phase as well as their abundance and location in the building vary greatly. In addition, these parameters depend on the type of material, the location (urban or rural environment, proximity to the sea, orientation in relation to the prevailing rains and winds, etc.) and environmental factors such as climate, atmospheric pollution or groundwater composition.

Salt crystallization inside porous materials depends on the nature of the solutions present in the pores and the conditions under which it occurs. Environmental, climatic and pollution conditions have changed in the past and continue to change today. In the past, the main factor that controlled the formation of salts in urban environments was air pollution. Although air pollution continues to be a parameter to be considered, its impact has lessened due to environmental policies that have reduced the concentrations of sulfur and other pollutants in the atmosphere (Brimblecombe et al. 2006).

Salt Crystallization in Porous Media,
coordinated by Hannelore DERLUYN and Marc PRAT.

In Europe, the directive (2016/2284/EU) "National Emissions reduction Commitments (NEC)"[1] has set pollutant emission limits for the various European countries, in particular for SO_2 and NO_x, thus contributing to a gradual reduction in the presence of these pollutants in the environment. For the European Union, the objective is to reduce emissions of SO_2 by 79% and NO_x by 63% between 2005 and 2030. In this air pollution reduction scenario, the role of climate change becomes the main factor conditioning the future behavior of salts in stones and building materials (Sabbioni et al. 2006).

The most common cations in building salts are Ca^{2+}, Na^+, Mg^{2+} and K^+, and the most common anions are CaO_3^{2-}, Cl^-, SO_4^{2-} and NO_3^-. Air pollution is not the only source of anions and cations that will interact with materials to form salts. The ions can have a natural (soil, marine aerosols) or anthropogenic origin (fertilizers, de-icing salt, pollution, interaction with other materials such as mortars) (Price 1996; Ruedrich and Siegesmund 2007). The composition of the salts will depend not only on current conditions but also on past ones: air quality during the building's history, materials used in construction, and during maintenance and restoration. For example, particles of wood burned in the 13th century are still present on the heads of the statues of the kings of Judah from Notre-Dame de Paris, which are currently in the Cluny National Museum of the Middle Ages in Paris (Del Monte et al. 2001). The effects of pollution on buildings can continue to be active for centuries after the pollution source has disappeared. Scrivano and Gaggero (2020), in their study on the formation of salts on buildings, highlighted the effect of chemical reactions between salt solutions, mainly sulfates from the dissolution of SO_2 in water, and rock minerals, mainly calcite.

The distribution of salt types in buildings is not uniform and depends on their solubility and the characteristics of the building materials (Charola 2000). Arnold (1982) differentiates four zones in buildings according to the solubility of the salts found. In the lowest part, we find the least soluble salts, mainly calcite and gypsum, then above this, a second zone with sodium nitrate and magnesium and sodium sulfates, which is the zone with the greatest deterioration. In the upper part, we find the most soluble salts such as nitrates and sodium chloride. Finally, the top zone corresponds to salt-free materials.

Depending on the presence of water in the liquid or vapor phase in the monuments, two main areas can be distinguished in which the precipitation/dissolution of salts occurs by different phenomena: on the one hand, the areas where the liquid water arrives by capillarity of the subsoil or surfaces exposed to rain, and, on the other hand, areas not directly affected by liquid water.

1 Available at: www.eur-lex.europa.eu.

In the first case, the water can transport elements coming from the basement or other parts of the building (mortars, coatings, metal elements, glass, etc.), which will interact with the stone, and at the same time the water can dissolve salts already present and transport them elsewhere (Franzen and Mirwald 2009). The extent of the wetted area may vary depending on the amount of water in the ground or with the frequency and intensity of rainy periods. The alternation of the arrival of liquid water and its subsequent disappearance by evaporation or drying will produce dissolution/precipitation cycles. In other areas, water will be brought mainly by moisture in the air. Dissolution will take place by deliquescence phenomena and the phase changes of the salts will depend on the temperature and relative humidity (RH) of the air.

This chapter has three main sections. First, a study of the composition of salts present in buildings is presented, emphasizing their variability in composition and in respect to environmental conditions. Then, we will see how these climate and air pollution conditions have changed over time and how atmospheric models predict their future evolution. Finally, these two types of information are compared in order to predict the way in which salt degradation can change over time depending on environmental conditions.

In this chapter, we are interested in the impact of climatic factors on the degradation of materials by salt crystallization and the evolution of this degradation over time in response to variations in environmental conditions. To be able to estimate the extent of degradation by salts, or more precisely its variation over time, either increase or decrease, it is necessary to establish quantitative relationships between environmental factors and parameters that characterize the degradation caused by salts. We will mainly consider the salts present in buildings made of stone, but this same analysis can be done for rock outcrops in natural environments or in other construction materials such as bricks and mortars.

8.2. Crystallization/dissolution and phase change of salts depending on environmental conditions

The damage produced by salts in porous materials has different origins (Kwaad 1970): (i) crystallization in a supersaturated solution, which results in a mixture of crystals and solutions occupying a larger volume than that of the original solution; (ii) absorption of water by salts to form hydrated salts; (iii) crystal growth; (iv) the thermal expansion of salts, generally greater than that of minerals. Buildings rarely contain a single type of salt, but rather a complex mixture of different ions that can result in the precipitation of a wide variety of salts (Arnold and Zehnder 1991).

8.2.1. *Behavior of simple salts*

To understand the degradation caused by salts within porous materials, the concept of crystallization pressure (Steiger 2005), presented in Chapter 2, is often used. If we are interested in the crystallization conditions and phase changes of salts present in buildings, we must consider their phase diagram. This will depend on three main factors: the concentration of salts in solution, the temperature and the RH. Let us take as an example two of the most studied salts that show two quite different behaviors: sodium sulfate and sodium chloride. These two salts represent in fact two extreme cases: sodium chloride with a single degree of hydration and a single polymorph at positive temperatures and sodium sulfate which presents much more complex crystallization processes, with phases having different degrees of hydration, some being metastable under natural conditions, with also polymorphs for the same degree of hydration (Steiger and Asmussen 2008).

Figure 8.1 shows the phase diagrams in the temperature–concentration or RH domain for these two salts. For temperatures between 0 and 30°C, the solubility of sodium chloride is almost constant (around 26% by mass) while that of sodium sulfate varies greatly (between 5% and 20% by mass). Temperature changes will more easily induce precipitation/dissolution for sodium sulfate than for sodium chloride. Similar conclusions can be drawn for RH changes in the case of crystals. In the case of sodium sulfate, the crystalline phase changes between the anhydrous salt (thenardite) and the decahydrate salt (mirabilite) and the deliquescence will depend on both temperature and RH. In the case of sodium chloride, with a single degree of hydration at temperatures above 0°C, deliquescence is almost independent of temperature, with an almost constant equilibrium RH around 75%. In general, salts with only one hydration phase have a lower degradation capacity than salts with several hydration phases.

With regard to the degradation caused by salts in building materials, not all salts react in the same way to changes in environmental conditions. Laboratory studies have shown that sodium sulfate causes degradation in stone when they are wetted with liquid water. On the other hand, materials contaminated with sodium chloride are much more affected by changes in RH (Shahidzadeh-Bonn et al. 2010; Menéndez and Petravona 2016).

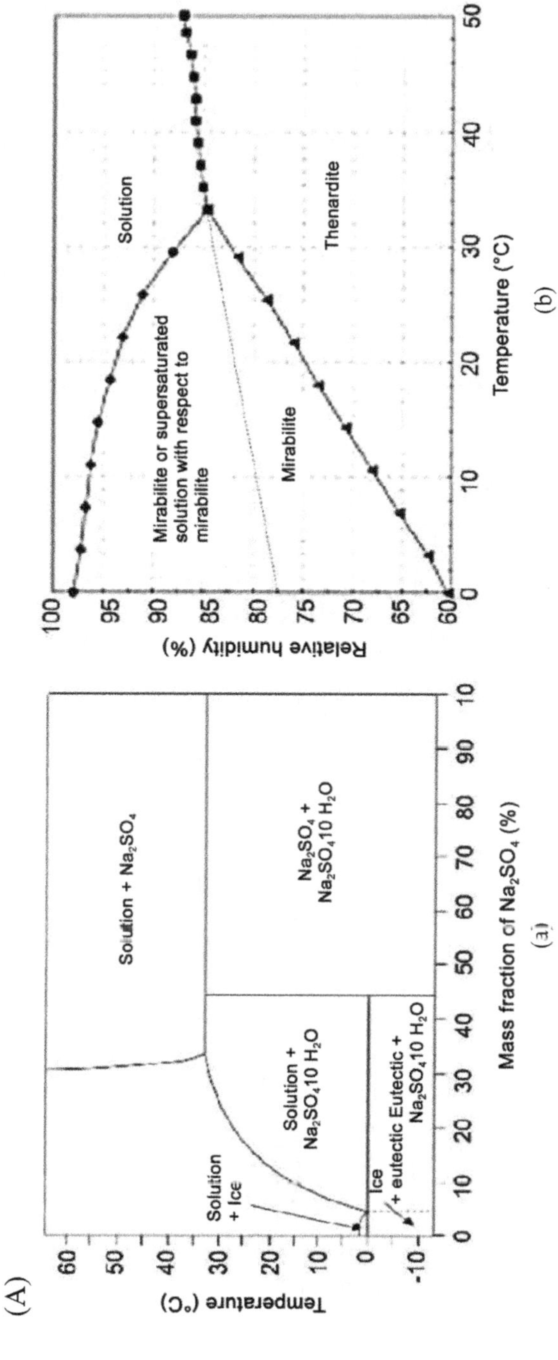
(A)
Temperature (°C)
Solution + Ice
Ice + eutectic
Eutectic + $Na_2SO_4$10 H_2O
Solution + $Na_2SO_4$10 H_2O
Solution + Na_2SO_4
Na_2SO_4 + $Na_2SO_4$10 H_2O
Mass fraction of Na_2SO_4 (%)
(a)
Relative humidity (%)
Solution
Mirabilite or supersaturated solution with respect to mirabilite
Mirabilite
Thenardite
Temperature (°C)
(b)

(B)

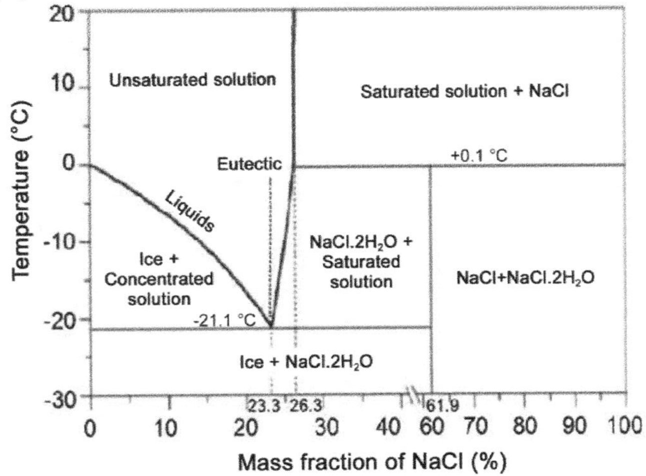

(a)

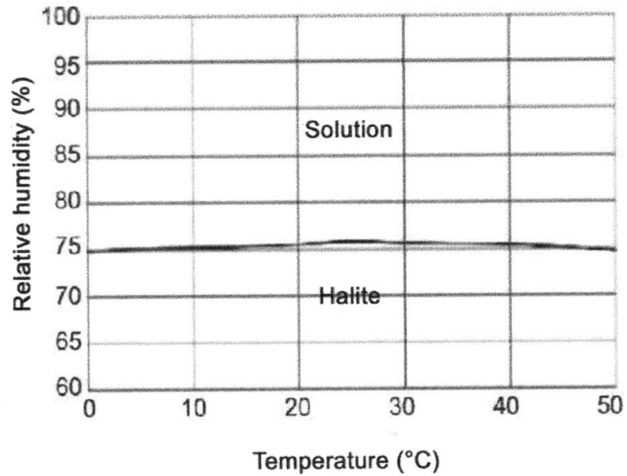

(b)

Figure 8.1. *(A) Sodium sulfate phase diagrams: (a) temperature/concentration; (b) temperature/relative humidity. (B) Sodium sulfate chloride for B phase diagrams: (a) temperature/concentration; (b) temperature/relative humidity*

8.2.2. *Behavior of complex salts*

In nature, as in buildings, it is common to find combinations of salts other than pure salts, which complicates the study of the dependence of salt crystallization on environmental conditions. There is a great variety of salts that can form by the crystallization of solutions containing several different anions and cations. In a study of sodium sulfate, its resources and its use, Wells (1923) describes various double or triple salts that can form by the precipitation of complex solutions. Figure 8.2 shows an example of the nature of salts precipitated from solutions containing sodium and magnesium sulfates as well as sodium and potassium chlorides at different temperatures.

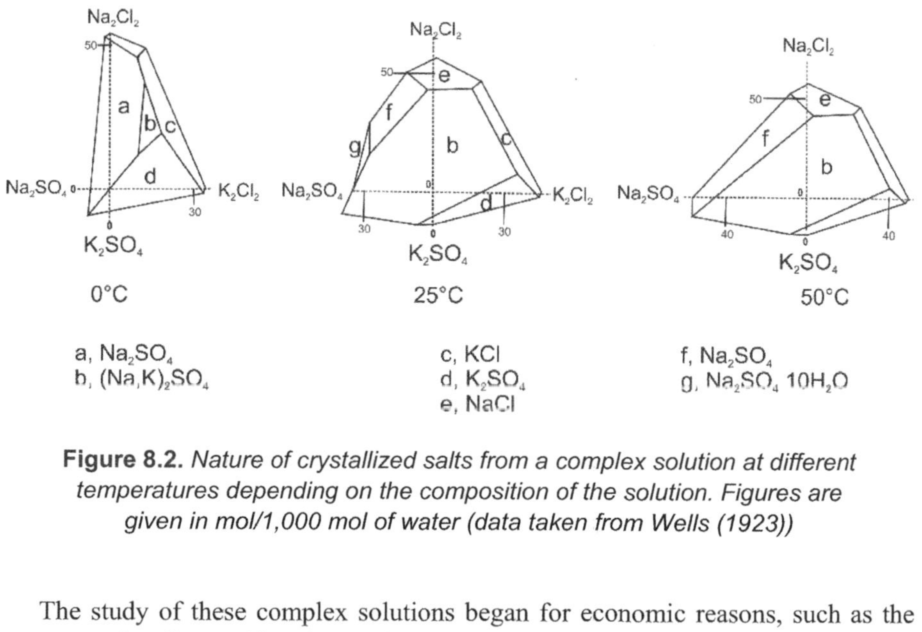

Figure 8.2. *Nature of crystallized salts from a complex solution at different temperatures depending on the composition of the solution. Figures are given in mol/1,000 mol of water (data taken from Wells (1923))*

The study of these complex solutions began for economic reasons, such as the recovery of sodium sulfate from saline solutions or the processes of desalination of sea water (Block and Waters 1968) and is now being researched again to understand the behavior of salts in porous materials. The behavior of mixed solutions is much more complicated than that of solutions containing a single species of cation and anion. For complex solutions, crystallization processes depending on composition and environmental conditions cannot be directly deduced from that of salts separately analyzed. An important parameter, such as solubility, can be greatly influenced by the presence of other ions. Consider the case of gypsum, an almost ubiquitous salt in the degradation products of buildings and which has very low solubility at room temperature, with a maximum of 2.1 g/L at 38°C. Gypsum in the presence of sodium chloride (NaCl) can multiply its solubility by four when the

NaCl concentration increases from 0 to 13% (Cameron 1901). Beyond 13% and at 40°C, the solubility of gypsum decreases to values lower than 5% when the NaCl concentration reaches saturation (Mockobey 1932). In the case of seawater, this solubility can be even greater due to the presence of magnesium (Furby et al. 1968). The presence of NaCl in the solution can also influence the balance between gypsum ($CaSO_4 \cdot 2H_2O$) and anhydrite ($CaSO_4$): the equilibrium temperature between the two phases is 40°C but it can drop to 14°C in a solution saturated with NaCl (Zen 1965). In Figure 8.3, we find the phases present in a complex solution of NaCl, Na_2SO_4 and $CaSO_4$. In this case, sodium chloride does not precipitate but thenardite (Na_2SO_4) precipitates at concentrations lower than those indicated in the phase diagram of sodium sulfate. Mirabilite ($Na_2SO_4 \cdot 10H_2O$) does not precipitate when the NaCl has a higher concentration than 20%. Gypsum, which has a solubility of 0.3%, precipitates at concentrations of 0.1% in the presence of the other salts.

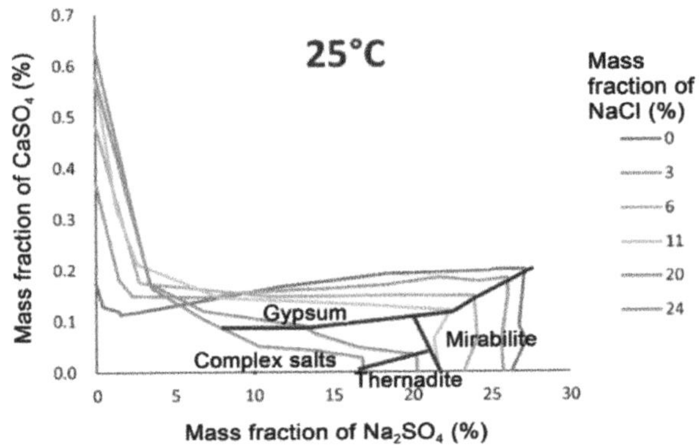

Figure 8.3. *Nature of crystallized salts (solid phase) in complex solutions at 25°C depending on the original composition of the solution. Complex salts, $Na_2Ca(SO_4)_2$+$CaSO_4 \cdot 2Na_2SO_4$-$2H_2O$, glauberite and "labile salt" (data from Block and Waters (1968)). For a color version of this figure, see www.iste.co.uk/derluyn/salt.zip*

Several authors have carried out laboratory tests to estimate the damage caused by combinations of different salts compared to single salts. We present here some examples of results obtained with different mixtures.

8.2.2.1. Na_2SO_4-$NaNO_3$

The behavior of the mixture Na_2SO_4-$NaNO_3$ was studied by De Clercq et al. (2013) and Lindström et al. (2016). In the study by Lindström et al. (2016), wetting/drying tests with water were carried out on samples of sandstone previously contaminated with simple and mixed saline solutions at concentrations around 10%

by mass. The samples were dried at an RH below 10%. The results showed that the observed degradation is greater in the samples contaminated only with sodium sulfate followed by those contaminated with the mixed solutions, and that those contaminated with sodium nitrate show the lowest degree of alteration. The explanation given by the authors was based on the fact that the crystallization pressure of mirabilite produced by the dissolution of pure sodium sulfate at a temperature of 23.5°C is 10 MPa, and only between 3.4 and 6.3 MPa for darapskite dissolution ($Na_3\ NO_3\ SO_4\ H_2O$). The authors also found that in the case of darapskite and sodium sulfate, the deformation capacity increases strongly when the temperature decreases.

De Clerq et al. (2013) also studied crystallization from simple and binary solutions at concentrations between 10% and 13.5% depending on the salt, but in two types of limestone, Maastricht and Euville, by evaporation at 20°C and at two RHs (50% and 85%) in samples contaminated with saline solutions. The porosities and drying rates of the two rocks are very different. The drying rate of Euville limestone with a porosity of 10% is approximately five times slower than that of Maastricht limestone with a porosity of 47.5%. On the other hand, in samples contaminated with saline solutions, Euville stone dries faster than Maastricht stone, perhaps due to its lower salt content. In the case of samples contaminated with an equimolar solution of sodium sulfate and sodium nitrate after 30 h of drying at room temperature and 50% RH, the authors found that, in the Maastricht stone, 10% of the salts introduced into the samples were found in the form of thenardite and darapskite efflorescence, and that, in the Euville stone, 43% of the salts initially present are found only as darapskite efflorescence. According to the authors, the difference can be explained by different crystallization pathways, associated with different evaporation rates. A faster evaporation rate, with slow mirabilite nucleation, will lead to saturation of darapskite, while a slower evaporation rate will allow saturation of mirabilite, which will transform into thenardite. In the case of drying at a RH of 85%, for both stones, the efflorescence is formed of thenardite with approximately 15% of nitratine ($NaNO_3$), which the authors attribute to precipitation after collection of the efflorescence. The distribution of salts between the surface and inside the samples is not the same in both types of stones and depends on their drying rate. To study the effect of changes in RH, samples dried at 50% RH were placed in an 85% environment. In this case, the samples containing the double solution present a greater degradation than the samples containing a single salt.

These studies show the complexity of the precipitation of incongruous salts (the salts used to form the solution are different from the precipitated salt), such as darapskite, whose deliquescence and crystallization phenomena are much more complex than for simple salts (Linnow et al. 2013). By studying the crystallization sequence of $NaNO_3$ and Na_2SO_4 solutions by Raman spectroscopy, these authors highlighted the precipitation of metastable phases and/or the inhibition of the precipitation of certain phases, with sequences and precipitations of salts different from those predicted by the diagram of theoretical phases.

8.2.2.2. Na_2SO_4-$NaCl$-$CaSO_4$

In another example studied by Menéndez and Petranova (2016), samples of three limestones (Lutetian (or Paris) stone, Caen and Vernon stones) were subjected to contamination with sodium chloride solutions (6% by mass), sodium sulfate (6% by mass), calcium sulfate (saturation) and a mixture of the three (6% by mass). The first contamination stage was carried out by capillarity and spraying during input and drying cycles. The salts were added at room temperature and the drying was carried out at two different temperatures and RHs: 20°C (RH 90% and 45%) and 50°C (RH 90% and 5%). During the wetting/drying cycles, and for all temperatures and RHs, the authors concluded that all stones, particularly Caen stone and Vernon chalk, were more strongly degraded by sodium sulfate than by the other solutions. In general, the cycles carried out at high RH cause less damage than those carried out at low RH, even if the authors observed an opposite effect for the Caen limestone, which has the smallest pore size. Cycles with drying at 50°C cause greater degradation than those at 20°C. After these cycles of contamination, the samples were subjected to RH cycles (2 weeks at 90%–2 weeks at 5% at 50°C and 2 weeks at 90%–2 weeks at 35% at 20°C), which showed that the salt with the greatest loss of mass was sodium chloride followed by the mixture of salts at 20°C and the mixture of salts at 50°C.

8.2.2.3. $MgSO_4$-$NaCl$-$CaSO_4$

Another salt commonly found in building materials is magnesium sulfate. Lopez Arce et al. (2008) carried out RH cycles (between 30% and 95%) at room temperature on samples of limestone saturated with a supersaturated solution of magnesium sulfate and a mixture of three salts obtained with solutions at concentrations of 50% mass for magnesium sulfate, 5% for sodium chloride and 27% for calcium sulfate. This study showed that the samples contaminated by the mixture of salts, after the RH cycles, present a higher and morphologically different degree of alteration than that of the samples contaminated only by the magnesium sulfate.

8.2.2.4. $MgSO_4$-Na_2SO_4-K_2SO_4

Cardell et al. (2008) carried out salt crystallization tests with a continuous supply at room temperature and 40% RH, using solutions of calcium (0.24% by mass), sodium (10%), potassium (11%), and magnesium (40%) sulfate as well as double solutions (50% mixture of single solutions) of calcium and magnesium sulfate, and sodium and magnesium sulfate. They concluded that the single solutions produce more damage than the double solutions in the case of tests at constant RH.

8.2.2.5. *$Mg(NO_3)_2$-$NaNO_3$-KNO_3-NaCl-$CaSO_4$*

Sawdy and Heritage (2007) studied the behavior of binary mixtures of salts: 16 g of nitromagnesite ($Mg(NO_3)_2$ H_2O, or 16 g of halite (NaCl) with 8 g of nitratine ($NaNO_3$), 4 g of niter (KNO_3) or 4 g of gypsum ($CaSO_4$ $2H_2O$). These salts were chosen because they cover a wide range in terms of solubility and equilibrium relative humidity, in addition to being generally present in painted structures. The study involved subjecting crystals of the different salts to variations in RH on either side of the equilibrium RH value of the salts studied. The authors also carried out evaporation tests in solutions saturated with mixtures of two salts. They found that the composition of the solution had a significant effect on the deliquescence and crystallization kinetics, which they explained mainly by the lower vapor pressure in a complex solution than in a solution with a single type of salt.

In general, this study leads to the conclusion that the rate of deliquescence/dissolution of a mixture of salts is faster than for a single salt, that the onset of deliquescence is controlled by the lowest equilibrium humidity between the two salts, and that the speed of evaporation of a solution with two salts is slower than for a single salt. For all these reasons, when the RH varies around the equilibrium RH, deliquescence is more likely in the presence of several salts, which usually occurs in buildings, rather than in the case of a single salt.

8.2.2.6. *Gypsum*

The role of gypsum in the deterioration of buildings is well known but difficult to simulate experimentally. The results of laboratory tests fail to reproduce the type of degradation that occurs on buildings, unlike other salts for which the degradations obtained by accelerated tests simulating natural processes are similar to those observed naturally. Gypsum crystallization tests in the laboratory do not, in most cases, produce significant degradations (Alves et al. 2021).

Menendez and David (2013) showed experimentally that under conditions of high RH gypsum precipitates within the material as opposed to surface crystallization under conditions of low RH. This difference can be interpreted by considering the fact that the evaporation takes place in two stages according to the model proposed by Hammecker (1995), which indicates that for salts where transport by molecular diffusion is not negligible, such as is the case of gypsum (Shahidzadeh et al. 2015), the concentration of the solution inside the materials increases with a uniform distribution and a slower drying, favoring precipitation within. Janvier-Badose et al. (2015) showed experimentally that the only possibility of finding gypsum at depths greater than 1 cm in the material is when the reaction of SO_2 in the vapor phase with water entering the pores occurs at this depth.

The presence of other salts in solution also contributes to degradation by gypsum (Goudie and Viles 1997). For McCabe et al. (2013), the diffusion rate is not the same in sandstones containing a solution of sodium chloride as when the solution also contains calcium sulfate. In a test involving both water and a solution of sodium chloride through a sample of sandstone, the authors observed that the diffusion inside the samples was slower when calcium sulfate was added to the sodium chloride.

8.2.2.7. *Complex salts in materials*

In light of previous studies, we see the significant effect of changes in RH on the degradation of materials containing mixtures of salts. The examples presented show us how the impact of salts on construction materials depends on (i) factors specific to the materials, mainly the characteristics of the porous systems; (ii) the composition of the salts; and (iii) environmental factors.

An important parameter that must be considered in the studies on the degradation of materials is the distribution of salts inside the materials, which depends on the factors mentioned above. Menendez and Petranova show that, in general, sodium chloride mainly precipitates near the surface while sodium sulfate can be found at a depth of more than 10 mm (Figure 8.4).

Similar results using different experimental procedures were found by Nunes et al. (2021). The localization of the crystals can also depend on the environmental conditions, mainly on the evaporation rate, which depends on the temperature and the RH. During slow evaporation, there is continuity between the water inside the stone and the surface, which allows salt transport to the surface. Here, advection plays an important role in the transport of salts such as sodium chloride (Shahidzadeh et al. 2015), whereas, during rapid evaporation, the only possibility of drying the samples is by diffusion of the water in the form of vapor, without transporting the salts, which remain within the stone. De Clercq et al. (2013) found that Euville and Maastricht stones contaminated with a solution of Na_2SO_4-K_2SO_4, after drying at 20°C and 50% RH, had a much greater distribution of ions at the surface than at depth.

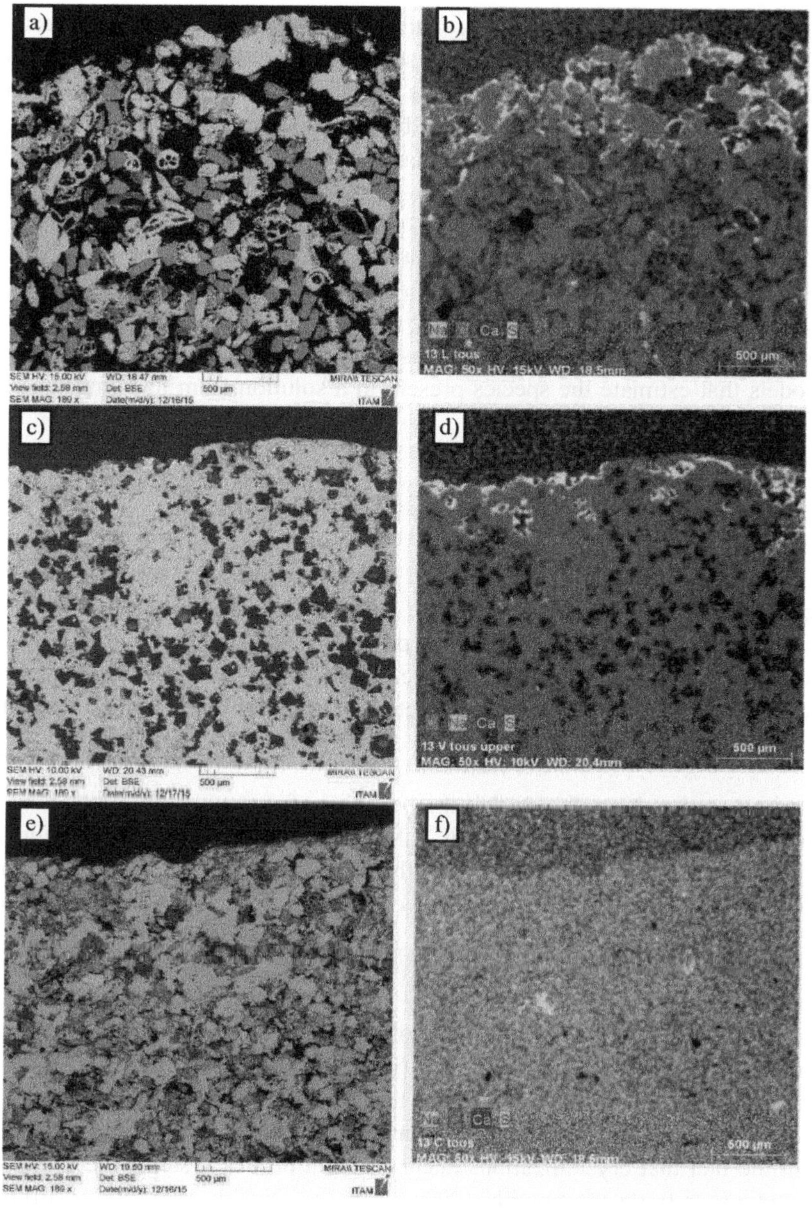

Figure 8.4. *Mapping of chemical elements in samples of Lutetian limestone (a,b), Vernon chalk (c,d) and Caen stone (e,f) with a mixture of sodium sulfate, sodium chloride and calcium sulfate. The chemical elements represented are green, Na; red, Cl; blue, Ca; light blue, S (images obtained at ITAM in Prague). For a color version of this figure, see www.iste.co.uk/derluyn/salt.zip*

As part of the SCOST project, Salt Compatibility of Surface Treatments (FP4-ENV 2C-ENV4980710), a series of salt concentration threshold values, below which salts do not cause degradation, has been established for different materials. These threshold values are no longer valid in the case of salt combinations and the behavior of the salts is more complex partly due to the formation of double salts (De Clercq 2008).

A complete compilation of the various articles focusing on the degradation of stones by different salts was provided by Alves et al. (2021). This work shows the great diversity of compositions and study conditions that have been tested and the difficulty of extracting general conclusions. One way to approach the problem is to use models that estimate the species present in a solution from its composition and environmental conditions.

8.2.3. *Modeling the crystallization of complex salts*

In the case of simple salt solutions, the crystallization sequence and phase changes can be deduced from phase diagrams according to the solution concentration, temperature, and RH. This procedure does not consider the dynamics of the processes which, as we have seen, have an impact on the sequence of crystallization with the formation of metastable phases or the inhibition of the nucleation of stable phases.

Previous studies have shown that the behavior of systems with mixtures of salts is much more complex than that of salts taken individually. It is difficult to empirically predict the nature of the salts that will crystallize and the effects that the variation in climatic conditions will have on phase changes and crystal growth (Steiger and Heritage 2012).

Thermodynamic models can be very useful for modeling the sequence and conditions of crystallization of salts in a solution, but their limitations must be taken into account (Steiger and Heritage 2012): (i) these models only consider equilibrium conditions, without the metastable phases that can form in particular conditions, and without the supersaturation phenomena that can take place; (ii) they do not take into account the fact that crystallizations occur inside a porous system where fluids can move and where fragmentation phenomena can occur inside micropores; (iii) model predictions are only valid if all ions present in the fluid are in the model.

The salts precipitated in complex solutions will also depend on the environmental conditions. Thermodynamic models have been used to study the precipitation of salts under confined or unconfined conditions. These models help to explain why some salts produce greater degradation in materials than others with higher crystallization pressures (Shahidzadeh-Bonn et al. 2010). The salts precipitated in complex solutions will also depend on the environmental conditions. Thermodynamic models have been used to study the precipitation of salts under confined or unconfined conditions. These models help to explain why some salts produce greater degradation in materials than others with higher crystallization pressures. Benavente et al. (2015) used this model to calculate the crystallization pressures of different salts (Na_2SO_4, $MgSO_4$, NaCl, Na_2CO_3, $CaSO_4$). Another model, among the most used in salt precipitation studies, is the thermodynamic model ECOS (Environmental Control Of Salts) (Price 2000) with its graphical interface RUNSALT (Bionda 2005). ECOS makes it possible to quickly determine the nature of the salts that will precipitate according to the concentrations, the temperature and the RH. This model also makes it possible to estimate the volume of salt precipitated from the environmental conditions in a solution with the cations and ions frequently found in the fluids present in buildings.

An example using the ECOS-RUNSALT model to calculate the evolution of the volume of salts as a function of temperature and RH conditions can be observed in Figure 8.5(a) for a solution composed of 1 mole of chloride, 2 moles of sulfate and 2 moles of nitrate, with 4 moles of sodium, 1 mole of magnesium and 1 mole of potassium. This theoretical composition was chosen arbitrarily to illustrate the results produced by the model. Figure 8.5(b) shows how calcium sulfate volume changes with environmental conditions. A limitation of this model is the control of gypsum in the crystallization pathway. In Figure 8.6, we can observe the volume variation for each phase present.

In the example shown in Figure 8.5, the volume of precipitated salts generally increases when the RH decreases and/or when the temperature increases. For this concrete case, the only conditions for which the volume of salt increases with an increase in RH or a decrease in temperature correspond to temperatures below 15°C at the time of the formation of mirabilite. Based on this type of modeling, Menéndez (2018) developed a methodology to define the temperature and RH conditions associated with increases in the volume of salts when the RH varies.

In a study on the Muñoz Chapel of the Cathedral of Cuenca (Spain), Martinez-Martinez et al. (2021) found two types of salts, sodium and magnesium sulfates. Using a methodology like that described by Menéndez (2018), two conditions were found under which damage can occur: (i) when RH increases, and anhydrite ($CaSO_4$) transforms into gypsum ($CaSO_4$ $2H_2O$) and starkeyite ($MgSO_4$ $4H_2O$) transforms into hexahydrite ($MgSO_4 \cdot 6H_2O$); (ii) when the RH decreases and the direct crystallization of $MgSO_4 \cdot 6H_2O$ occurs in the pores.

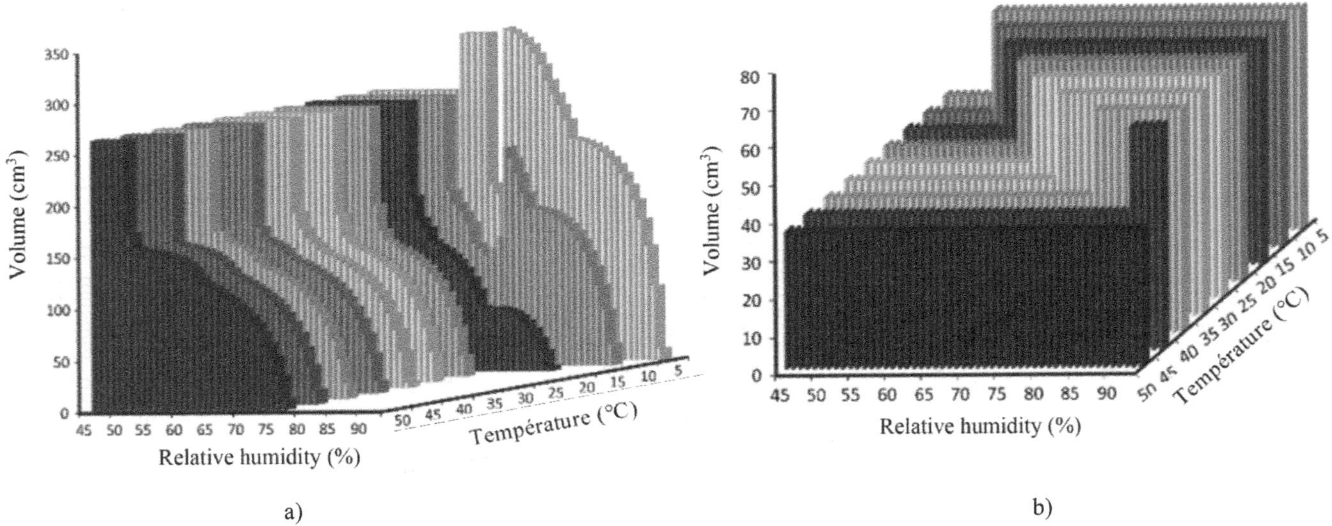

Figure 8.5. *Variation of the total volume of precipitated salts as a function of temperature and relative humidity: (a) for a solution of 1 mole of Cl^-, 2 moles of NO_3^-, 2 moles of SO_4^{2-} and 4 moles of Na^+, 1 mole of Mg^{2+} and 1 mole of K^+; (b) same as for a solution of 1 mole of Ca^{2+} and 1 mole of SO_4^{2-}. These volumes were calculated using the model ECOS-RUNSALT. For a color version of this figure, see www.iste.co.uk/derluyn/salt.zip*

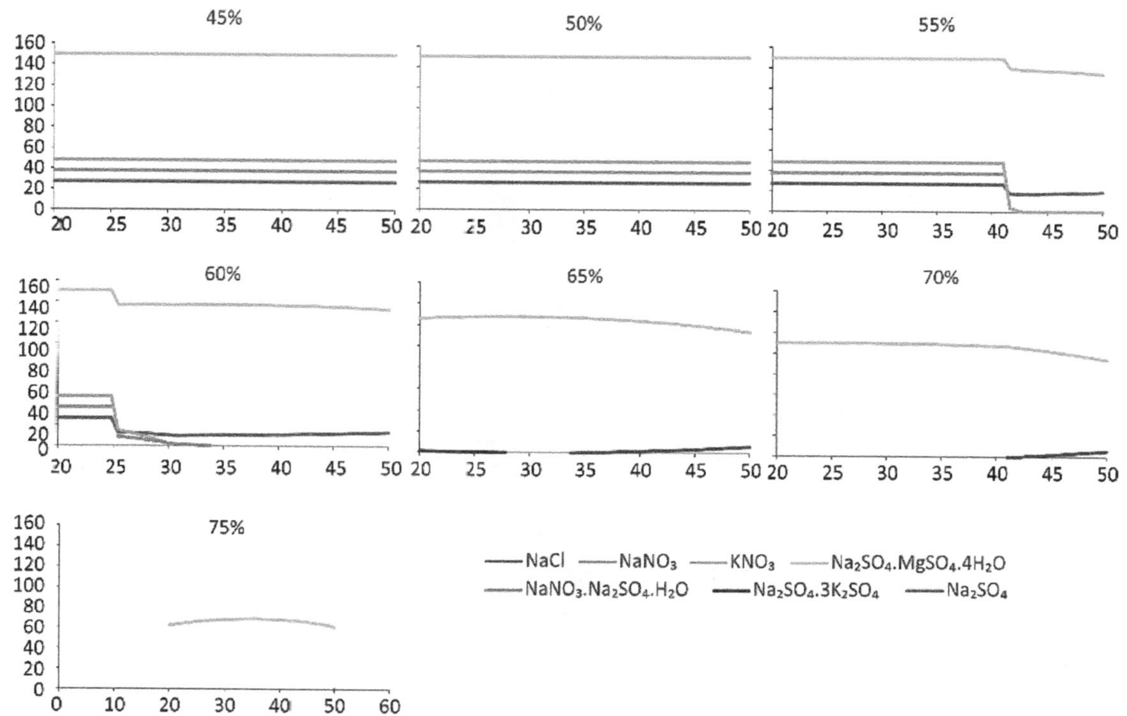

Figure 8.6. *Same data as in Figure 8.5 but with the separation of the volumes of each phase. The vertical axis expresses the volume in cm^3 and the horizontal axis expresses the temperature in °C for different relative humidities indicated at the top of the figures. For a color version of this figure, see www.iste.co.uk/derluyn/salt.zip*

The advantage of this methodology is that it can be applied to very different compositions, if the ions present are among those considered by the RUNSALT application. These data can be combined with meteorological or climatic data in order to estimate the damage produced by salts in materials. The results obtained will be simplified estimates because the complexity of actual cases is difficult, if not impossible, to integrate into the calculations.

8.2.4. *Model limits*

An important factor is the reaction kinetics that conditions the time necessary to make the dissolution and crystallization processes operational. Colston et al. (2001), in a study carried out at St Andrew's Church in Walpole (Norfolk, United Kingdom), where the meteorological data of the site and the variation in the surface condition of the materials over time were recorded, found that the degradation produced by sodium chloride is greater during periods when the RH is below 75% for several hours, between 400 and 600 h, than when the cyclical variations around 75% RH are more frequent, thus corresponding to shorter periods (Figure 8.7).

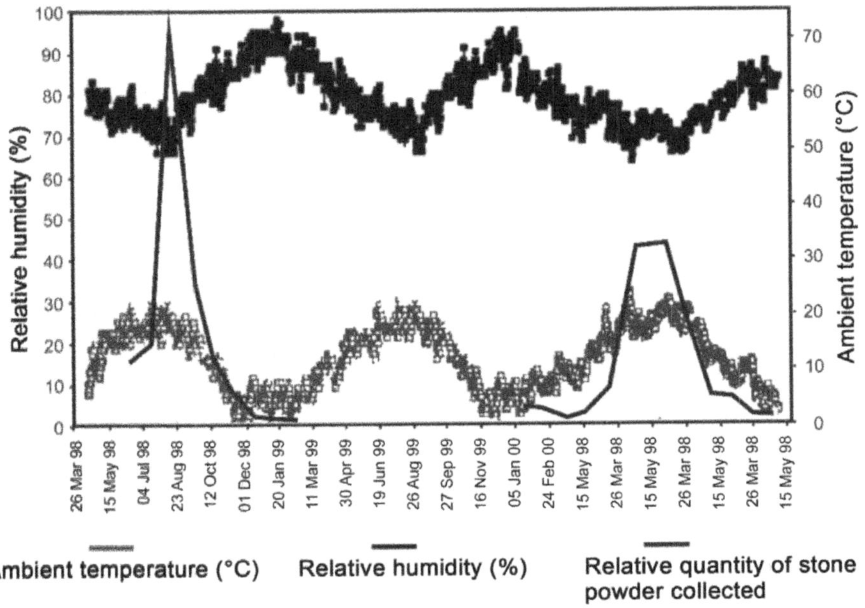

Figure 8.7. *Comparison between temperature and relative humidity conditions inside St Andrew's Church and the loss of material collected (data from Colston et al. (2001)). For a color version of this figure, see www.iste.co.uk/derluyn/salt.zip*

Godts et al. (2021) studied the crystallization and deliquescence phenomena of NaCl inside glass capillary tubes with an internal diameter of 0.1 mm. They found that the time required for the dissolution of the crystals when the RH varies between 20% and 90% is 22 min in the capillaries where the crystals are close to the extremity of the capillary, and between 68 and 144 min in capillaries where the crystal is placed farther from the end, between 0.37 and 0.47 mm. By considering the effect of the distance from the tip and the volume of solution at the time crystallization begins, a direct relationship can be established between the distance from the opening of the capillary and the time required for crystallization or the dissolution of salts in the capillaries. For the dissolution experiments, crystals are found at 20 μm, 0 μm, 362 μm, and 411 μm from the end of the capillary with decay times between 22 min and 2 h and 24 min. The authors compared these values with those obtained for drops of solution on a glass plate. In the latter case, and for the same difference in RH (between 90% and 20%), they obtain crystallization times of 8 min and 25 s, and 5 min and 44 s for dissolution.

These results show that the experiments made in unconfined environments and the theoretical simulations do not consider the complexity of the phenomena that take place in actual cases in buildings. It has been shown that the evaporation kinetics can have a direct effect on the crystallization sequence, with the precipitation of metastable phases or with the inhibition of the nucleation of certain phases. When modeling the crystallization/dissolution and hydration/dehydration processes, for salts with different degrees of hydration, it is also necessary to consider the duration of the process, which is not the same in a pore (capillary) than in a solution in the open air.

8.3. Change in environmental conditions

So far, we have seen how salt crystallization strongly depends on the environmental conditions in which it occurs. We will now study how these conditions have changed over time and how this information can be used to predict the impact of environmental changes on the future behavior of salts in materials. Climatic conditions have changed naturally since the formation of the Earth and will continue to do so. We have, for example, large cycles of ~100,000 years that occurred during the past 800,000 years, or cycles of ~500 years such as medieval global warming or the Little Ice Age between the 14th and 19th centuries. Effects caused by humans must be added to this natural variability, since man contributed greatly to climate change that began last century, which is ongoing and will continue in the future.

Knowledge of future conditions is of paramount importance for preventive conservation. The use of conservation materials and methods adapted to the future climate can extend the durability of current restorations and avoid the use of materials that will no longer be suitable for future conditions.

The term "heritage climatology" has been used to understand the long-term degradation of monuments from the work carried out in the NOAH's ARK project (Brimblecombe 2010). If climatology is the science that studies average meteorology over time, heritage climatology can be defined as the study of climatological parameters that affect historical monuments, materials and sites. Subsequently, this research methodology was applied more particularly to the action of salts with the name "salt climatology" (Grossi et al. 2011). According to Brimblecombe (2010), heritage climatology is more concerned with the conditions in the materials than in the air and the parameters that must be considered are temperature, rainfall, RH, and wind speed and direction. These are the parameters that control the degradation of materials. Even if for built heritage one might think that the change of a few degrees will not have a significant impact on the behavior of materials, this is not always the case. For example, the corrosion of certain metals is maximum between 9 and 11°C because the interaction with liquid water is significant (Kuzera et al. 2007) and therefore even a very small change in temperature can have a great impact on the corrosion of these metals. The freeze/thaw processes in the pores, which result in volume variations generating tensions inside the pores, occur at temperatures close to 0°C and can lead to significant degradation. Increasing temperatures in the future may reduce the number of freeze/thaw cycles in some regions of the world, but it may also increase them, for example, in buried archaeological remains in Greenland. In these regions, the soils have so far been prevented from thawing but a small increase in temperature could lead to an increase in the number of freeze/thaw cycles (Grossi et al. 2007). Similarly, the equilibrium conditions of the different salts are highly dependent on environmental parameters, mainly temperature and RH. These conditions will also influence the rate of evaporation of saline solutions and therefore the variation in concentration of fluids inside the pores, as well as their spatial distribution within the materials. The combination of all these parameters will condition the degradation power of the salts.

The parameters most often considered to estimate the degradation of materials are temperature and RH. Other parameters such as rainfall, sunshine duration, and wind direction and speed are not directly considered in most studies on salts, even if they are important on a small scale and on the building scale.

8.3.1. *Atmospheric pollution*

Salts in buildings have three main origins: (i) the incorporation of atmospheric pollutants into the materials; (ii) the reactions of these materials with products resulting from the leaching of other materials used in construction such as mortars or coatings; (iii) the effect of pollutants contained in groundwater (phytosanitary products, de-icing salts, products resulting from accidental pollution). We will deal in this chapter only with the case of atmospheric pollution, which is one of the main factors in the formation of salts currently present in buildings.

To understand the current state of conservation of the materials, it is important to know the conditions in which the building evolved. This knowledge will allow us to estimate how the building will behave in the future with changing conditions. Several authors have been interested in the study of past air pollution, mainly in cities. An example of this type of study is presented by Brimblecombe (1977) on air pollution in London from 1500 to 1900. Estimating the amount of SO_2 and smoke in the air is made from the consumption of coal in the city. Using a similar methodology, Ionescu et al. (2012) and Brimblecombe and Lefevre (2021) estimated the concentrations of atmospheric pollutants in Paris since the year 1300 with an extrapolation up to 2100 (Figure 8.8). In this figure, recent data were obtained from the Hygiene Laboratory of the City of Paris and AIRPARIF, and future data were estimated with the model GAINS (The Greenhouse Gas and Air Pollution Interactions and Synergies).

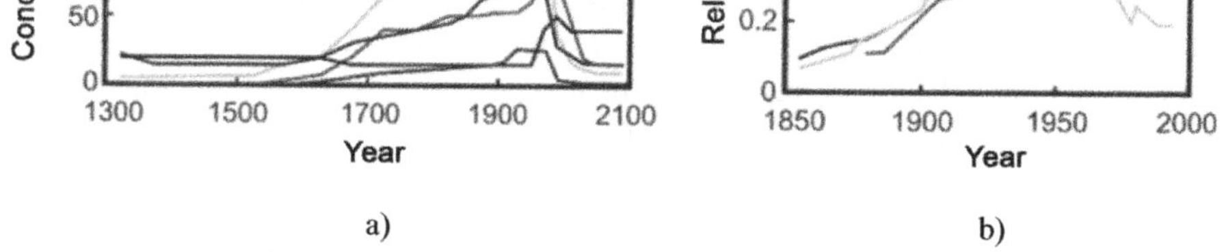

a) b)

Figure 8.8. *(a) Concentration of SO_2, NO_2, O_3, PM_{10} (solid particles with a diameter less than 10 µm), EC (elemental carbon) in Paris from 1300 to 2100 (data from Brimblecombe and Lefevre (2021)). (b) Deposition of pollutants in central Germany (Schellenberger Costa et al. 2022). For a color version of this figure, see www.iste.co.uk/derluyn/salt.zip*

As shown in Figure 8.8(a), the concentrations of different pollutants in Paris, mainly sulfur dioxide, increased steadily from the Middle Ages until the 2000s, with a significant acceleration corresponding to the invention of the steam engine and the use of coal at the end of the 18th century. After the middle of the 19th century, the rate of SO_2 in the atmosphere in Paris decreased due to the replacement of coal by gas and electricity for heating. Similar data have been presented by other authors for other geographical regions such as pollutant deposition in central Germany shown in Figure 8.8(b) (Schellenberger Costa et al. 2022). In the latter case, the decrease in the deposition of pollutants comes a little later, probably due to different choices in the energy policy between the two countries.

Different models allow us to make predictions about the composition of the atmosphere in the future. For example, here we used the model GAINS, an extension of RAINS which is a model of pollutant emissions and their impacts. GAINS incorporates the main greenhouse gases (CO_2, CH_4, N_2O) and fluorinated gases (HFC, PFC, SF6, and NF3). In the simulations in Figure 8.9, the ECLIPSE v6 CLE scenario was used: this is a reference scenario that considers future environmental policies similar to current policies. Simulations of variations in sulfur and nitrogen deposition for Europe are shown in Figure 8.9. A larger reduction in sulfur than in nitrogen is observed.

Concentrations of atmospheric pollutants have been used as an input parameter in dose–response functions, which make it possible to quantify the reaction of materials to external effects. For example, we present the function for the erosion of limestone developed by Lipfert in 1989 (Bonazza et al. 2009):

$$L = 18.8 \cdot R + 0.016 \cdot [H^+] \cdot R + 0.18 \cdot (V_{dS}\,[SO_2] + V_{dN}\,[HNO_3])$$

where L = decrease in thickness (μm/year); 18.8 = constant (μm/m) based on the solubility of calcite at a concentration of CO_2 at 330 ppm; R = precipitation (m/year); 0.016 = constant valid for the precipitation of calcite at a pH between 3 and 5; $[H^+]$ = hydrogen ion concentration (μmol/l) calculated from annual rainfall; 0.18 = conversion factor of (cm/s) (μg/m^3) in μm; V_{dS} = deposition rate of SO_2 (cm/s); $[SO_2]$ = concentration of SO_2 (μg/m^3); V_{dN} = deposition rate of HNO_3 (cm/s); $[HNO_3]$ = concentration of HNO_3 (μg/m^3).

This same type of function can be established for other phenomena such as the corrosion of metals or the degradation of glass. For the formation of salts, there is no such function because the processes are difficult to model. On the other hand, it is easy to understand that future changes in the quantities of SO_2 or NOx in the atmosphere will result in a reduction in the formation of salts in buildings.

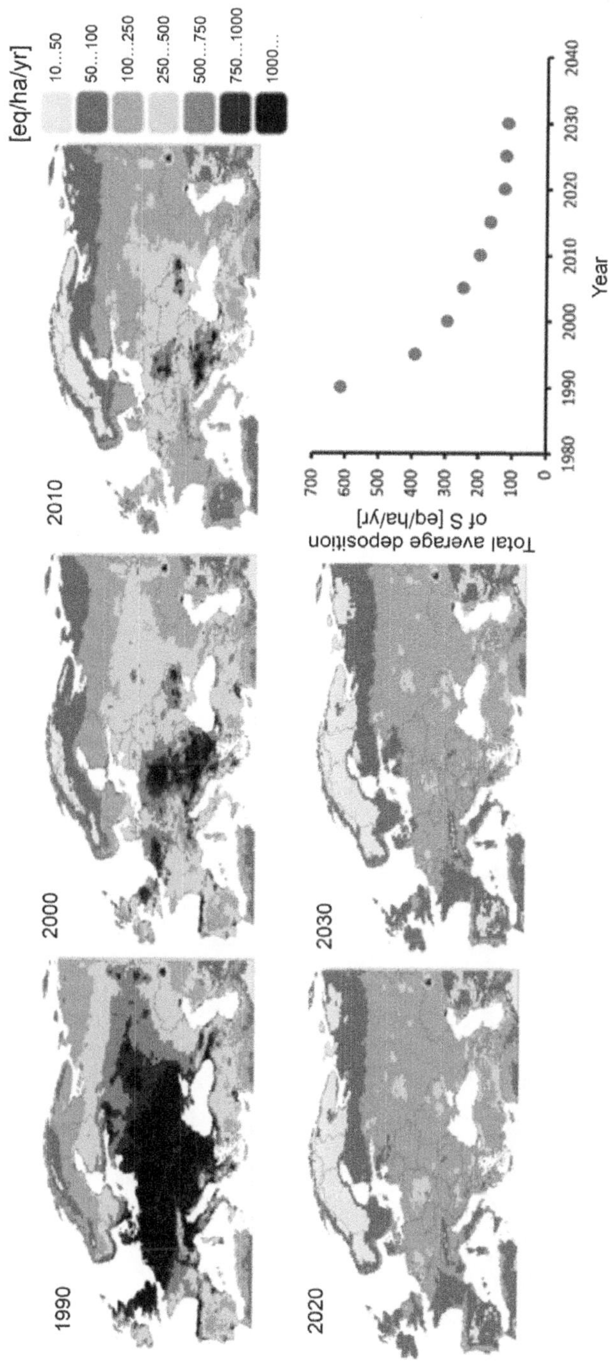
1990
2000
2010
2020
2030
[eq/ha/yr]
10...50
50...100
100...250
250...500
500...750
750...1000
1000...
Total average deposition of S [eq/ha/yr]
Year

(a)

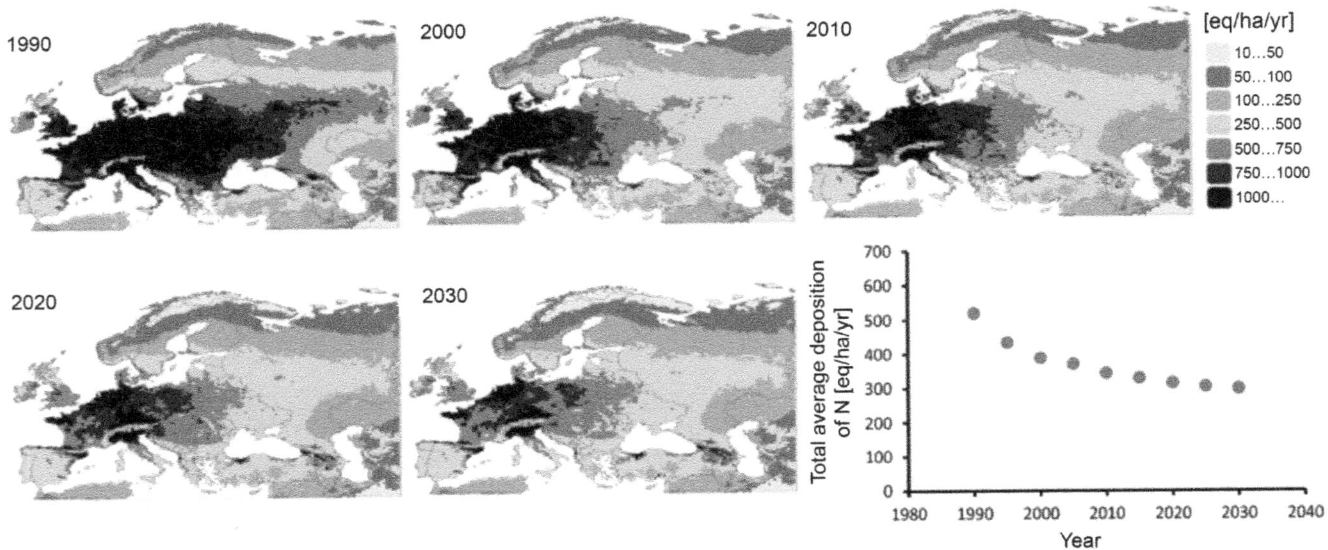

(b)

Figure 8.9. *(a) Variations in the deposition of S and N. (b) Variations obtained by the GAINS model between 1990 and 2030 in Europe. For a color version of this figure, see www.iste.co.uk/derluyn/salt.zip*

8.3.2. *Climate*

Atmospheric pollution is the “source” of ions that forms salts, and weather conditions will control the formation and dissolution of these salts. We will now study how climatic conditions are expected to vary over time. It is important to consider two aspects of climate change: a slow and continuous change in parameters (warming, decrease in RH) and a change in the frequency and magnitude of extreme events. To illustrate these changes, we will focus on France.

Many climate models use the emission scenarios proposed by the IPCC (Intergovernmental Panel on Climate Change) to predict the future climate. These scenarios are called representative concentration pathways (RCP). There are four scenarios, RCP 8.5, RCP 6.0, RCP 4.5, RCP 2.6, from the most pessimistic, RCP 8.5, where nothing changes and greenhouse gas emissions continue to increase at the current rate, to the most optimistic scenario, RCP 2.6, very low emissions scenario with a peak before 2050. Scenarios 6.0 and 4.5 correspond to situations with a stabilization of emissions before the end of the 21st century at a medium (6.0) or low (4.5) rate. Climate models have similar general trends but there may be some differences in the simulations. To eliminate these differences, multi-model simulations that consider the results of a set of individual models can be used. It is this approach that we used to illustrate climatic variations in France, with the DRIAS portal of Météo France, which provides all users with simulations from different models for reference time intervals (1951–2005), near future (2021–2050), middle future (2041–2070) and far future (2071–2100), and for different scenarios (RCP 2.6, 4.5, 8.5).

In the example in Figure 8.10, scenario 4.5 has been chosen, with a policy that aims to stabilize CO_2 emissions. In this figure, two parameters that can have a strong impact on the behavior of the salts are taken into account: rainfall for the dissolution and hydration of the salts, and the number of days of heat waves (temperatures of at least 5°C greater than normal for at least 5 consecutive days) for the precipitation and dehydration of salts. This figure shows the benefit of carefully selecting the parameters for analysis. Variations in mean values, such as precipitation, are smaller than those for the extreme values, which can have a significant effect on salt crystallization processes, as demonstrated by Colston et al. (2001) with the duration of drought periods.

The last step in predicting the future behavior of salts in buildings is to find the relationship between salt type and atmospheric conditions.

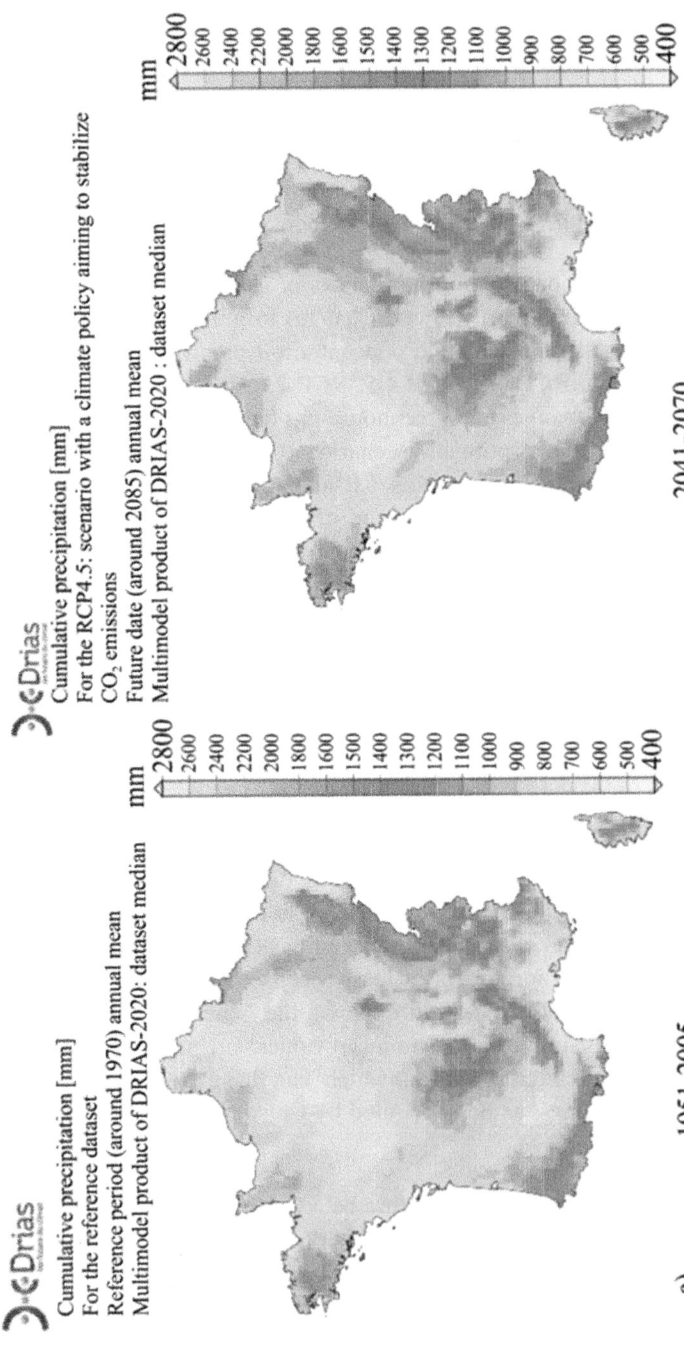
Drias
Cumulative precipitation [mm]
For the reference dataset
Reference period (around 1970) annual mean
Multimodel product of DRIAS-2020: dataset median
1951-2005
a)
mm
2800
2600
2400
2200
2000
1800
1600
1500
1400
1300
1200
1100
1000
900
800
700
600
500
400
Drias
Cumulative precipitation [mm]
For the RCP4.5: scenario with a climate policy aiming to stabilize
CO_2 emissions
Future date (around 2085) annual mean
Multimodel product of DRIAS-2020 : dataset median
2041-2070
mm
2800
2600
2400
2200
2000
1800
1600
1500
1400
1300
1200
1100
1000
900
800
700
600
500
400

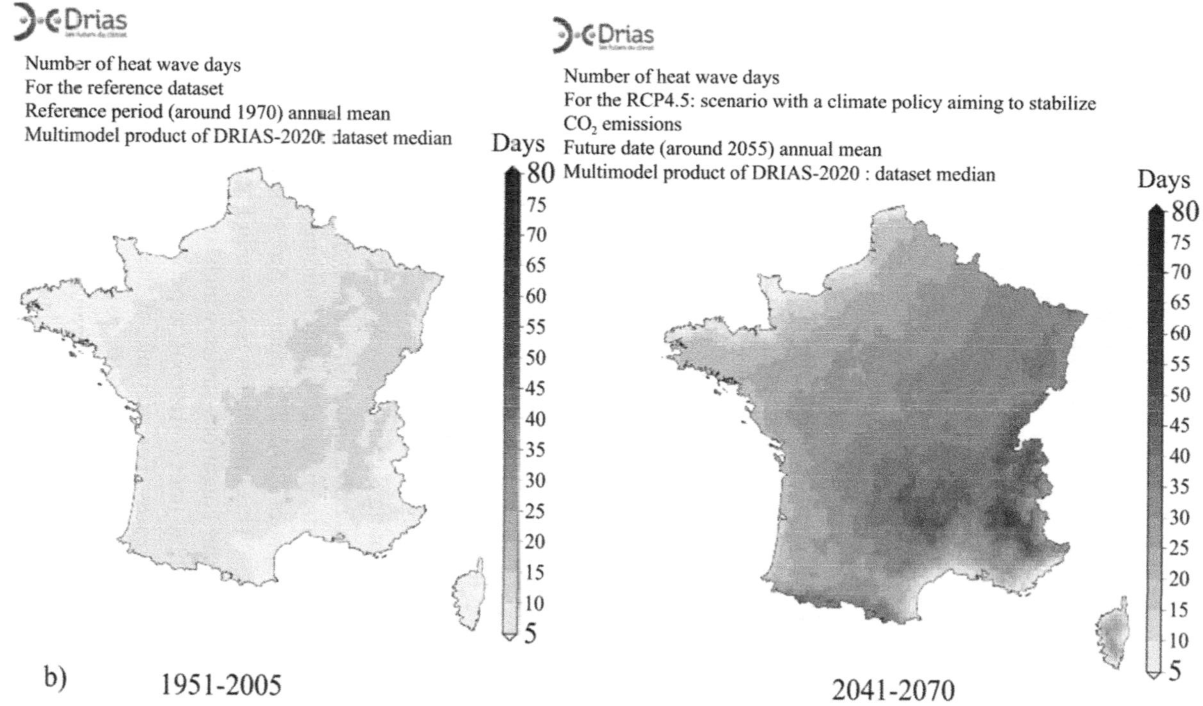

Figure 8.10. *(a) Cumulative precipitation. Data from the DRIAS site of Météo France corresponding to the average of data from different models. (b) Number of heat wave days. Data from the DRIAS Météo France corresponding to the data mean from different models. For a color version of this figure, see www.iste.co.uk/derluyn/salt.zip*

8.4. Estimation of future degradation of materials by salts

8.4.1. *Methods used*

Several methods have been used to estimate the damage created by different salts depending on the climatic conditions. These methods do not consider the material itself, only the crystallization/dissolution or hydration/dehydration processes that may occur. There are two main method types. In the first group, we count the number of times in the phase diagram where the environmental conditions cross the line that separates the liquid phase from the solid phase, or in other words a lower state of hydration and a higher state. Let us examine the two most studied cases. For sodium chloride, we will count the number of times the RH drops from more than 75.3% (Grossi et al. 2011; Menéndez 2018) to less than 75.3%, a limit that corresponds to crystal precipitation. This limit varies according to the author, some take the value 65.5%, which corresponds to the RH of NaCl deliquescence in a mixed solution, instead of 75.3%, (Godts et al. 2021). For sodium sulfate, the thenardite/mirabilite transformation limit is not constant for all temperatures but it can be adjusted easily by the equation: RH (%) = 0.72 T (°C) + 60.4 between 0 and 32°C (Menéndez 2018). In this case, the transformation of thenardite into mirabilite occurs when the RH increases. A similar equation was used by Benavente et al. (2008) for temperatures below 22.5°C and which corresponds to the conditions where the crystallization pressure is above 10 MPa.

We have seen that most materials do not contain a single salt but a mixture of salts and that the variation in the volume of crystallized salts can be estimated with the ECOS model and the RUNSALT application. Menéndez (2018) used this type of simulation to define the conditions of temperature variation and RH for which the volume of salt increases, separating the situations when the RH increases or when it decreases. In this chapter, a complex solution corresponding to the anions and cations found in a sample of mortar from a Parisian building was used: 29.7 ppm of Cl^-, 105 ppm of SO_4^{-2}, 25.3 ppm of NO_3^-, 22.2 ppm of Na^+, 8.6 ppm of K^+, 4.9 ppm of Mg^{2+} and 96.9 ppm of Ca^{2+}. This information is shown in Figure 8.11 along with the equations that delineate the conditions under which an increase in volume is expected. The areas between blue lines correspond to periods when the RH increases and in red when the RH decreases. For example, if the RH increases and if it is between $-\ 0.36 \times$ temperature $+\ 63$ and $0.01\ \times$ temperature $+\ 70$, it can be considered that there will be an increase in volume. In the case where the temperature is below 7.5°C and the RH decreases, there are three areas in which the volume increases: (i) if $0.38 \times T + 47 <= RH <= 0.48 \times T + 50$; (ii) if $1.12 \times 7 + 41$

$<=$ RH $<= 1.14 \times T + 44$ (these conditions correspond to a range of temperatures between 0°C and 40°C; (iii) if $0.01 \times T^2 + 0.39 \times T + 16 <=$ RH $<= 0.01 \times T^2 + 0.39 \times T + 18$.

Menéndez (2017) compared these conditions with meteorological data from seven Météo France stations. An example for the city of Paris is shown in Figure 8.11. Average daily temperature and RH data between 2007 and 2013 were used for the calculation.

The same methodology was used in another case on meteorological data from 42 meteorological stations uniformly distributed over France, between 1971 and 2000, and simulations obtained with the Aladin model of Météo France and the RCP 4.5 scenario for the years 2071–2100 at the same geographical locations as the stations. In Figure 8.12, we can observe how the number of days when the volume of salt increases per year varies between the past and the future. In this map, there is no clear trend, but it can be seen that, in coastal areas, the extent of degradation remains constant.

In this type of study, we must not consider only the average variations per year because we know that in the future, we expect greater seasonal climatic variations with greater differences in average temperatures in winter than in summer, and greater differences in average RH in spring and fall. Precipitation will increase in winter but decrease in summer. Grossi et al. (2011) calculated, for the different seasons, the number of transitions for sodium chloride and sodium sulfate in Spain and observed a change in seasonality with, for NaCl, an increase in the number of transitions in fall and a decrease in summer. For Na_2SO_4, they obtain an increase in transitions in spring and fall and a decrease in summer.

Different authors (Colston et al. 2001; Grossi et al. 2011; Godts et al. 2021) found that transitions are generally more frequent in summer months and less frequent in winter for rural environments. Godts et al. (2021) established that the RH is lower in urban environments than in rural environments, with RHs closer to the sodium chloride crystallization threshold (75%) during winter in cities, which has as result in greater damage in winter than in summer, unlike in rural areas.

Another factor to consider is the fact that the increase in rain in winter implies a longer saturation of the materials, which can promote the migration of ions inside and the crystallization of salts at greater depths than in the past (McCabe et al. 2013).

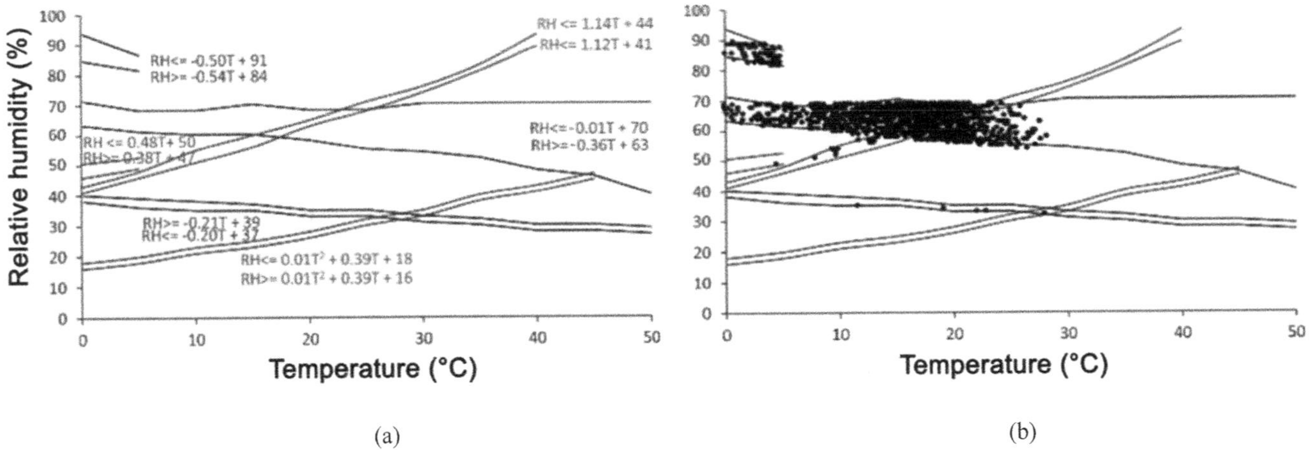

Figure 8.11. *(a) Conditions for which an increase in the volume of salts can occur when the relative humidity increases (in blue) or decreases (in red). Example (b) shows the days for which these conditions are met. Paris weather station between 2007 and 2013. For a color version of this figure, see www.iste.co.uk/derluyn/salt.zip*

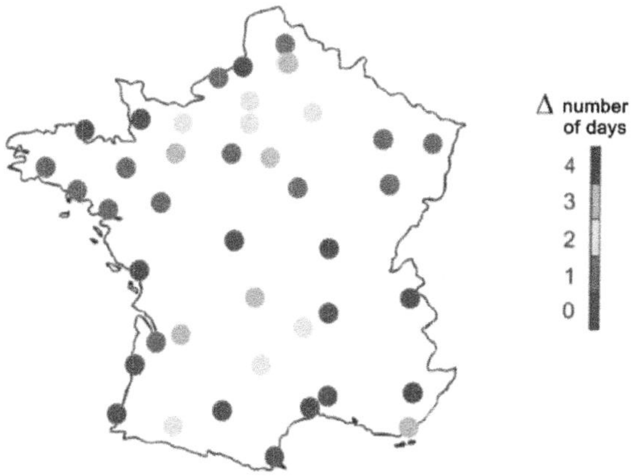

Figure 8.12. *Variation in the number of days when the volume of salts increases between 1971–2000 and 2070–2100. For a color version of this figure, see www.iste.co.uk/derluyn/salt.zip*

8.4.2. *Limitations of methods used*

These methods have several limitations because they do not consider, for example, the fact that not all crystallization/dissolution cycles necessarily result in damage. Indeed, when the cycles occur on the surface, they generate efflorescence, much less dangerous than sub- or crypto-efflorescence (Watt 1999).

Furthermore, it must also be considered that the temperature and RH can vary greatly with the orientation of the exposed surface and with the depth within the material. McAllister et al. (2016) carried out an important experimental study involving measuring the temperature and the RH at different depths in sandstone exposed in four directions of space (NE, SE, SW, NW). Thanks to this work, important checks have been made on generally accepted ideas. Environmental conditions have cycles of different durations. With regard to the crystallization/dissolution processes and phase changes of salts, we generally consider the diurnal cycles and sometimes also those that occur within a day. McAllister et al. (2016) concluded the following for the same type of material:

– Temperature fluctuations are greater at the surface than at depth. The temperature difference between the surface and the interior increases the flow of humidity, which can mobilize salts and cause them to precipitate in another place.

– RH cycles are also more frequent at the surface. The energy required to extract water is greater than for humidification. Drying time is longer than wetting time.

– The orientation of the exposed surface is important because, depending on the sunshine and the direction of the wind, it affects the temperature and the RH.

– The wind has a very strong impact on humidification and drying phenomena. During periods of rain, the wind can allow water to penetrate to significant depths in the material. During drying periods, it can produce rapid drying of the surface and promote a water break between the surface and the interior of the material.

In a case study by McAllister et al. (2016), the SO exposed surface undergoes the maximum number of wetting/drying cycles at depths between 5 and 50 mm due to the greater contribution of water by the driving rain in this direction and also because of the sunshine.

All of these factors make studies aimed at relating salt crystallization to meteorological data complex, and simplifications often have to be made. If the goal is to compare past behavior with the future, we can consider that these effects are equivalent for all the time periods considered. We compare the climatic data and we consider that the other parameters (orientation, characteristics of the materials, etc.) remain constant.

André et al. (2012) recorded the temperature and RH in two similar temples located a few hundred meters apart, in the Angkor region of Cambodia. These temples are located in the Angkor Archaeological Park with a monsoon climate, where 88% of the 1,320 mm of annual rainfall falls between May and October. One of the temples, Ta Keo, is in an area where vegetation was cleared in the 1920s and the other, Ta Nei, is still in an overgrown environment. Surface temperature and RH data were recorded from December 2008 to November 2009 at different locations in both temples. Table 8.1 shows the temperature and RH values for each temple. Differences in the data between the two locations are evident, in particular, the maximum temperature difference between the two temples is 10°C during the dry period and 12°C during the rainy period, with higher values in the temple cleaned of its vegetation. Regarding the minimum RH, it is two times lower in Ta Keo than in Ta Nei throughout the year. The variation in temperature and RH in a day are greater, almost double, in Ta Keo temple compared to Ta Nei.

Observation period/parameter		Ta Keo (without vegetation)	Ta Nei (invaded by vegetation)
Year (26 December 2008–15 November 2009)			
Surface temperature (°C)	**Maximum**	45.62	35.21
	Minimum	20.15	17.72
	Mean	30.89	27.06
Relative humidity (%)	**Maximum**	100	100
	Minimum	14.32	36.6
	Mean	73.37	85.16
Diurnal temperature variation (°C)	**Maximum**	20.97	7.5
	Minimum	2	0.5
	Mean	9.5	4.13
Variation in diurnal relative humidity (%)	**Maximum**	57.61	40.1
	Minimum	3.82	1
	Mean	32.59	17.46
Dry season (26 December 2008–30 April 2009)			
Surface temperature (°C)	**Maximum**	45.62	35.21
	Minimum	20.15	17.72
	Mean	32.25	27.08
Relative humidity (%)	**Maximum**	96.5	100
	Minimum	14.32	36.6
	Mean	55.99	74.17
Diurnal temperature variation (°C)	**Maximum**	20.97	7.5
	Minimum	2	1.5
	Mean	12.46	5.2
Variation in diurnal relative humidity (%)	**Maximum**	57.57	40.1
	Minimum	14.01	7.3
	Mean	38.18	24.06

Observation period/parameter		Ta Keo (without vegetation)	Ta Nei (invaded by vegetation)
Rainy season (1 May 2009–15 November 2009)			
Surface temperature (°C)	**Maximum**	43.63	31.72
	Minimum	23.15	22.72
	Mean	30.02	27.04
Relative humidity (%)	**Maximum**	100	100
	Minimum	32.91	61.9
	Mean	84.4	92.13
Diurnal temperature variation (°C)	**Maximum**	14.48	5.5
	Minimum	2.5	0.5
	Mean	7.24	3.31
Variation in diurnal relative humidity (%)	**Maximum**	57.61	34
	Minimum	3.82	1
	Mean	28.33	12.44

Table 8.1. *Meteorological data in two temples of the archaeological site of Angkor (according to André et al. (2012))*

During one week in the dry season, more precise measurements were made in the four orientations of space and for both temples. The authors found that the diurnal temperature differences depend not only on the vegetation cover of the site but also on the orientation of the walls. The temperature range is similar in the two buildings for the N facade (4°C); for the O and E facades, it is greater in Ta Keo (11°C) than in Ta Nei (5°C); on the contrary in the S facade, there is 11°C difference in Ta Keo and 17°C in Ta Nei. For RH, there are also differences depending on the sites and the orientation of the facades. The highest RH values are found in the N face of Ta Keo (91%) and in the E face of Ta Nei (97%) but the lowest RH, for the two sites, corresponds to the S façade with 17% for Ta Keo and 43% for Ta Nei.

These measurements highlight the well-known role of vegetation as a regulator of temperature and RH, but also of the effect of the orientation of the facades. The variations observed between the different facades are not the same for both buildings. Regarding the modeling of the effect of climate on salt crystallization, this study highlights the importance of local factors that are not considered in large-scale modeling from climate data.

Studies concerning salt crystallization in relation to the interior microclimate are less frequent than those on exterior facades (Martinez-Martinez et al. 2021) but one way to decrease salt damage is to modify interior conditions to reduce the number of crystallization/dissolution cycles (Cataldo et al. 2005).

8.5. Conclusion

If we wanted to sum up in one word the problem of salt crystallization in a changing environment, we would choose the word *variability*.

Variability in salt composition: as we have seen in buildings, we rarely have a single type of salt but a heterogeneous composition that varies from building to building, and within the same building it also varies depending on the location: interior or exterior, height, exposure of the façade to wind and/or sun, distance to the surface, etc.

Variability in the type of support: salts are found in any type of support, stone, bricks, mortar, coatings, etc. In each category of materials, the properties are very different in terms of chemical and mineralogical composition, porosity, grain size, pore size, etc.

Variability in climatic conditions and their evolution over time and space: climate on Earth is very varied depending on the geographical location: latitude, longitude, distance to the sea, altitude. These climates have changed in the past due to natural causes, but we must now add anthropogenic effects that cause changes over a very short time scale. There are two types of variations in weather conditions: slow and gradual variations, and variations in the frequency and magnitude of extreme events.

Variability in environmental conditions over time: a good illustration of this is the variation in atmospheric pollution over time, which is linked to socioeconomic development affecting the type of heating and mode of transport, policies to reduce greenhouse gases, the use of phytosanitary products in agriculture and gardens and the use of de-icing salts.

Variability in local environmental conditions: on a smaller scale, in a building, air pollution will affect the materials of the facades on the street-facing side and the garden-facing side differently, on a street with heavy traffic or on a pedestrianized street, on a facade facing the sun or the wind or protected from the sun or wind, etc. Similarly, the temperature and RH can vary by several degrees depending on the surrounding vegetation, sun exposure, wind and rain direction, etc.

Variability in the type of climatic phenomena: climate change not only has an impact on the slow and continuous variations in meteorological factors but also on the scale and frequency of extreme events such as heat waves, storms, periods of drought and strong rain. We must not neglect the secondary effects of these extreme events, which can produce, for example, floods. These floods can cause the transfer of a large number of pollutants, which will remain in buildings for a long time.

Variability in the kinetics of evaporation and dissolution phenomena: these can lead to crystallization paths different from those predicted by phase diagrams and thermodynamic models.

Even if with such a degree of variability we would tend to say predicting the behavior of salts according to environmental conditions is too complex to be modeled, it is always possible to make simplifications to average these effects and manage to establish general trends based on climate model simulations.

It is important to find out what the future environmental conditions will be, particularly if it may lead to the degradation of materials by salts or other weathering agents in order to be able to predict and act on conservation policies adapted to these new situations. This knowledge is based on the study of the past behavior of materials in buildings and on laboratory studies that simulate the observed changes in situ. It is essential to know how the materials will react to future environmental conditions because the restoration and conservation methods used differ depending on the climate and atmospheric pollution. It is just as important not to neglect human impact, which can change the environment of buildings through multiple actions such as modifying the level of sunshine on the walls by the construction of nearby buildings, by the planting of trees or creating green spaces, through work on buildings (restoration, water drainage, insulation), or finally through the collection of water, which affects groundwater levels.

8.6. References

Alves, C., Figueiredo, C.A.M., Sanjurjo-Sánchez, J., Hernández, A.C. (2021). Salt weathering of natural stone: A review of comparative laboratory studies. *Heritage*, 4(3) 1554–1565.

André, M.F., Voldoire, O., Roussel, E., Vautier, F., Phalip, B., Peou, H. (2012). Contrasting weathering and climate regimes in forested and cleared sandstone temples of the Angkor region. *Earth Surf. Process. Landforms*, 37, 519–532.

Arnold, A. (1982). Rising damp and saline minerals. *4th International Congress on the Deterioration and Preservation of Stone Objects*, Louisville.

Arnold, A. and Zehnder, K. (1991). *Monitoring Wall Paintings Affected by Soluble Salts*. The Getty Conservation Institute, London.

Benavente, D., Brimblecombe, P., Grossi, C. (2008). Salt weathering and climate change. In *New Trends in Analytical, Environmental and Cultural Heritage Chemistry*, Tassi, L. and Colombini, M.P. (eds). Transworld Research Network, Trivandrum.

Benavente, D., Brimblecombe, P., Grossi, C.M. (2015). Thermodynamic calculations for the salt crystallisation damage in porous built heritage using PHREEQC. *Environ. Earth Sci.*, 74, 2297–2313.

Bionda, D. (2005). RUNSALT computer program, 2002–2005 [Online]. Available at: http://science.sdf-eu.org/runsalt/ [Accessed 15 November 2021].

Block, J. and Waters, O. (1968). The $CaSO_4$-Na_2SO_4-$NaCl$-H_2O system at 25° to 100°C. *Journal of Chemical and Engineering Data*, 13(3), 336–344.

Bonazza, A., Messina, P., Sabbioni, C., Grossi, C.M., Brimblecombe, P. (2009). Mapping the impact of climate change on surface recession of carbonate buildings in Europe. *Science of the Total Environment*, 407, 2039–2050.

Brimblecombe, P. (1977). London air pollution, 1500-1900. *Atmospheric Environment*, 11, 1157–1162.

Brimblecombe, P. (2010). Heritage climatology. In *Climate Change and Cultural Heritage*, Lefèvre, R.-A. and Sabbini, C. (eds). Edipuglia, Bari.

Brimblecombe, P. and Lefèvre, R.A. (2021). Weathering of materials at Notre-Dame from changes in air pollution and climate in Paris, 1325–2090. *Journal of Cultural Heritage*, 50, 88–94.

Brimblecombe, P., Grossi Sampedro, C.M., Harris, I. (2006). Climate change critical to cultural heritage. *The International Conference on Heritage, Weathering and Conservation*, 1, 387–393.

Cameron, F. (1901). Solubility of gypsum in aqueous solutions of sodium chloride. *Journal of Physical Chemistry*, 5(8), 556–576.

Cardell, C., Benavente, D., Rodríguez-Gordillo, J. (2008). Weathering of limestone building material by mixed sulfate solutions. Characterization of stone microstructure, reaction products and decay forms. *Materials Characterization*, 59, 1371–1385.

Cataldo, R., De Donno, A., De Nunzio, G., Leucci, G., Nuzzo, L., Siviero, S. (2005). Integrated methods for analysis of deterioration of cultural heritage: The Crypt of "Cattedrale di Otranto". *Journal of Cultural Heritage*, 6, 29–38.

Charola, A.E. (2000). Salts in the deterioration of porous materials: An overview. *Journal of the American Institute for Conservation*, 39, 327–343.

Colston, B., Watt, D., Munro, H. (2001). Environmentally-induced stone decay: The cumulative effects of crystallization-hydration cycles on a Lincolnshire oopelsparite limestone. *Journal of Cultural Heritage*, 4, 297–307.

De Clercq, H. (2008). Performance of limestone contaminated with binary mixtures of sodium sulphate and treated with a water repellent. In *Hydrophobe V: 5th International Conference on Water Repellent Treatment of Building Materials*. Aedificatio Publishers, Rome.

De Clercq, H., Jovanovic, M., Linnow, K., Steiger, M. (2013). Performance of limestones laden with mixed salt solutions of Na_2SO_4–$NaNO_3$ and Na_2SO_4–K_2SO_4. *Environ. Earth Sci.*, 69, 1751–1761.

Del Monte, M., Aussetb, P., Lefevre, R.A., St' Thiebault, S. (2001). Evidence of pre-industrial air pollution from the Heads of the Kings of Juda statues from Notre Dame Cathedral in Paris. *The Science of the Total Environment*, 273, 101–109.

Franzen, C. and Mirwald, P.W. (2009). Moisture sorption behaviour of salt mixtures in porous stone. *Chemie der Erde–Geochemistry*, 69(1), 91–98.

Furby, E., Glueckauf, E., McDonald, A. (1968). The solubility of calcium sulphate in sodium chloride and sea salt solutions. *Desalination*, 4(2), 264–276.

Godts, S., Orr, S.A., Desarnaud, J., Steiger, M., Wilhelm, K., De Clercq, H., Cnudde, V., De Kock, T. (2021). NaCl-related weathering of stone: The importance of kinetics and salt mixtures in environmental risk assessment. *Heritage Science*, 9(1) [Online]. Available at: https://heritagesciencejournal.springeropen.com/articles/10.1186/s40494-021-00514-3.

Goudie, A. and Viles, H. (1997). *Salt Weathering Hazards*. Wiley, New York.

Grossi, C.M., Brimblecombe, P., Harris, I. (2007). Predicting long term freeze-thaw risk on Europe built heritage and archaeological sites in a changing climate. *Sci. Total Environ.*, 377, 273–281.

Grossi, C.M., Brimblecombe, P., Menéndez, B., Benavente, D., Harris, I., Déque, M. (2011). Climatology of salt transitions and implications for stone weathering. *Science of Total Environment*, 409, 2577–2585.

Hammecker, C. (1995). The importance of the petrophysical properties and external factors in the stone decay on monuments. *PAGEOPH*, 145, 338–361.

Ionescu, A., Lefèvre, R.A., Brimblecombe, P., Grossi, C.M. (2012). Long-term damage to glass in Paris in a changing environment. *Science of the Total Environment*, 431, 151–156.

Janvier-Badosa, S., Beck, K., Brunetaud, X., Guirimand-Dufour, A., Al-Mukhtar, M. (2015). Gypsum and spalling decay mechanism of tuffeau limestone. *Environ. Earth Sci.*, 74, 2209–2221.

Kucera, V., Tidblad, J., Kreislova, K., Knotkova, D., Faller, M., Reiss, D., Snethlage, R., Yates, T., Henriksen, J., Schreiner, M. et al. (2007). UN/ECEICP materials dose-response functions for the multi-polluants situation. *Water, Air, & Soil Pollution, Focus*, 7, (1-3), 249–258.

Kwaad, F.J.P.M. (1970). Experiments on the granular disintegration of granite by salt action. *Fysisch Geografischen Bodemkundig Laboratorium from Field to Laboratory*, 16, 67–80.

Lindstrom, N., Talreja, T., Linnow, H., Stahlbuhk, A., Steiger, M. (2016). Crystallization behavior of Na_2SO_4–$MgSO_4$ salt mixtures in sandstone and comparison to single salt behaviour. *Applied Geochemistry*, 69, 50–70.

Linnow, K., Steiger, M., Lemster, C., De Clercq, H., Jovanovic, M. (2013). In situ Raman observation of the crystallization in $NaNO_3$–Na_2SO_4–H_2O solution droplets. *Environ. Earth Sci.*, 69, 1609–1620.

Lipfert, F.W. (1989). Atmospheric damage to calcareous stones: Comparison and reconciliation of recent experimental findings. *Atmos. Environ.*, 23, 415–429.

Lopez-Arce, P., Doehne, E., Martin, W., Pinchin, S. (2008). Sales de sulfato magnésico y materiales de edificios históricos: simulación experimental de laminaciones en calizas mediante ciclos de humedad relativa y cristalización de sales. *Materiales de Construcción*, 58(289/290), 125–142.

Martínez-Martínez, J., Torrero, E., Sanz, D., Navarro, V. (2021). Salt crystallization dynamics in indoor environments: Stone weathering in the Muñoz Chapel of the Cathedral of Santa María (Cuenca, central Spain). *Journal of Cultural Heritage*, 47, 123–132.

McAllister, D., Warke, P., McCabe, S. (2016). Stone temperature and moisture variability under temperate environmental conditions: Implications for sandstone weathering. *Geomorphology*, 280, 137–152.

McCabe, S., Smith, B.J., McAlister, J.J., Gomez-Heras, M., McAllister, D., Warke, P.A., Curran, J.M., Basheer, P.A.M. (2013). Changing climate, changing process: Implications for salt transportation and weathering within building sandstones in the UK. *Environ. Earth Sci.*, 69, 1225–1235.

Menéndez, B. (2017). Estimation of salt mixture damage on built cultural heritage from environmental conditions using ECOS-RUNSALT model. *Journal of Cultural Heritage*, 24, 22–30.

Menéndez, B. (2018). Estimators of the impact of climate change in salt weathering of cultural heritage. *Geosciences*, 8(11), 401.

Menéndez, B. and David, C. (2012). Using acoustic emissions to estimate the environmental conditions controlling weathering by gypsum crystallization in a porous limestone. *Environ. Earth Sci.*, 68, 1691–1706.

Menéndez, B. and Petráňová, V. (2016). Effect of mixed vs single brine composition on salt weathering in porous carbonate building stones for different environmental conditions. *Engineering Geology*, 210, 124–139.

Mockobey, E.W. (1932). The solubility of calcium sulfate in sodium chloride solutions. Master's thesis, Missouri University of Sciences Technology [Online]. Available at: https://scholarsmine.mst.edu/masters_theses/4818 [Accessed 15 November 2021].

Nunes, C., Aguilar Sanchez, A.M., Godts, S., Gulotta, D., Ioannou, I., Lubelli, B., Menéndez, B., Shahidzadeh, N., Slízková, Z., Theodoridou, M. (2021). Experimental research on salt contamination procedures and methods for assessment of the salt distribution. *Construction and Building Materials*, 298, 123862.

Price, C.A. (1996). *Stone Conservation: An Overview of Current Research.* Getty Institute, Santa Monica.

Price, C.A. (2000). *An Expert Chemical Model for Determining the Environmental Conditions Needed to Prevent Salt Damage in Porous Materials.* Archetype Publications, London.

Ruedrich, J. and Siegesmund, S. (2007). Salt crystallisation in porous sandstone. *Environmental Geology*, 52, 225–249.

Sabbioni, C., Cassar, M., Brimblecombe, P., Tidblad, J., Kozlowski, R., Drdacky, M., Saiz-Jumenez, C., Grontoft, T., Wainwright, I., Arino, X. (2006). Global climate change impact on built heritage and cultural landscapes. In *Heritage, Weathering and Conservation*, Fort, R., Alvarez de Buergo, M., Gomez-Heras, M., Vazques-Calvo, C. (eds). Taylor & Francis Group, London.

Sawdy, A. and Heritage, A. (2007). Evaluating the influence of mixture composition on the kinetics of salt damage in wall paintings using time lapse video imaging with direct data annotation. *Environmental Geology*, 52, 303–315.

Schellenberger Costa, D., Ottoa, J., Chmara, I., Bernhardt-Römermann, M. (2022). Estimating historic N- and S-deposition with publicly available data – An example from Central Germany. *Environmental Pollution*, 292, 118378.

Scrivano, S. and Gaggero, L. (2020). An experimental investigation into the salt-weathering susceptibility of building limestones. *Rock Mechanics and Rock Engineering*, 53, 5329–5343.

Shahidzadeh-Bonn, N., Desarnaud, J., Bertrand, F., Chateau, X., Bonn, D. (2010). Damage in porous media due to salt crystallization. *Phys. Rev. E*, 81, 066110.

Shahidzadeh-Bonn, N., Schut, M.F.L., Desarnaud, J., Prat, M., Bonn, D. (2015). Salt stains from evaporating droplets. *Scientific Reports*, 5, 10335.

Steiger, M. (2005). Crystal growth in porous materials I: The crystallization pressure of large crystals. *Journal of Crystal Growth*, 282, 455–469.

Steiger, M. and Asmussen, S. (2008). Crystallization of sodium sulfate phases in porous materials: The phase diagram Na_2SO_4–H_2O and the generation of stress. *Geochimica et Cosmochimica Acta*, 72, 4291–4306.

Steiger, M. and Heritage, A. (2012). Modelling the crystallisation behaviour of mixed salt systems: Input data requirements. In *12th International Congress on the Deterioration and Conservation of Stone.* Columbia University, New York.

Watt, D.S. (1999). *Building Pathology: Principles and Practice*, 2nd edition. Blackwell, Hoboken.

Wells, R.C. (1923). Sodium sulphate: Its sources and uses. *United States Geological Survey Bulletin*, 717, 43 [Online]. Available at: https://www.google.com/url?sa=t&rct=j&q=&esrc=s&source=web&cd=&cad=rja&uact=8&ved=2ahUKEwit7c2h64mBAxVwRKQEHYGFCDQQFnoECA4QAQ&url=https%3A%2F%2Fpubs.usgs.gov%2Fbul%2F0717%2Freport.pdf&usg=AOvVaw22B5EXGluqq6FsnXbfjRU_&opi=89978449.

Zen, A.N. (1965). Solubility measurements in the system CaS04-NaCl-HzO at 35°, 50°, arid 70° C and one atmosphere pressure. *Journal of Petrology*, 6, 124–164.

List of Authors

Eduardo E. ALONSO
Technical University of Catalonia
Barcelona
Spain

Patrick DANGLA
Navier, CNRS
École des Ponts ParisTech
Université Gustave Eiffel
Marne-la-Vallée
France

Hannelore DERLUYN
LFCR, CNRS
and
DMEX, CNRS
Université de Pau et des Pays de l'Adour
France

Julie DESARNAUD
Labo des monuments et
décoration monumentale
KIK-IRPA
and
CSTC
Brussels
Belgium

Beatriz MENÉNDEZ
GEC
CY Cergy Paris Université
Neuville-sur-Oise
France

Lionel MERCURY
ISTO, CNRS
BRGM
Université d'Orléans
France

Marc PRAT
IMFT, CNRS
Université de Toulouse
France

Anna RAMON-TARRAGONA
Technical University of Catalonia
Barcelona
Spain

Noushine SHAHIDZADEH
IoP
Van der Waals-Zeeman Institute
University of Amsterdam
The Netherlands

Index

E, F

G, H

I, K, M

T, W, X

List of Authors

Eduardo E. ALONSO
Technical University of Catalonia
Barcelona
Spain

Patrick DANGLA
Navier, CNRS
École des Ponts ParisTech
Université Gustave Eiffel
Marne-la-Vallée
France

Hannelore DERLUYN
LFCR, CNRS
and
DMEX, CNRS
Université de Pau et des Pays de l'Adour
France

Julie DESARNAUD
Labo des monuments et
décoration monumentale
KIK-IRPA
and
CSTC
Brussels
Belgium

Beatriz MENÉNDEZ
GEC
CY Cergy Paris Université
Neuville-sur-Oise
France

Lionel MERCURY
ISTO, CNRS
BRGM
Université d'Orléans
France

Marc PRAT
IMFT, CNRS
Université de Toulouse
France

Anna RAMON-TARRAGONA
Technical University of Catalonia
Barcelona
Spain

Noushine SHAHIDZADEH
IoP
Van der Waals-Zeeman Institute
University of Amsterdam
The Netherlands

Printed and bound by CPI Group (UK) Ltd, Croydon, CR0 4YY

13/09/2026

14960998-0004